WILD LIFE IN
A SOUTHERN COUNTY

Richard Jefferies

Illustrated by
C.F. Tunnicliffe

LITTLE TOLLER BOOKS
an imprint of THE DOVECOTE PRESS

This paperback edition published in 2011 by
Little Toller Books
Stanbridge, Wimborne Minster, Dorset BH21 4JD
First published in 1879 by Smith, Elder & Co.

ISBN 978-1-908213-00-6

Introduction © Richard Mabey 2011
Illustrations © The Estate of C.F. Tunnicliffe 2011

Typeset in Monotype Sabon by Little Toller Books
Printed in Spain by GraphyCems, Navarra

All papers used by Little Toller Books and the Dovecote Press
are natural, recyclable products made from
wood grown in sustainable, well-managed forests

A CIP catalogue record for this book is available
from the British Library

1 3 5 7 9 8 6 4 2

CONTENTS

INTRODUCTION

Richard Mabey

RICHARD JEFFERIES' *Wild Life in a Southern County* was my first encounter with nature writing. I was about twelve years old, and quite content with my rag-bag collection of I-Spy books and bird guides, texts about what was what, and where to find it. When I found my elder sister's copy of *Wild Life*, I was mesmerised. Here were thoughts about how animals might think, and how landscapes made you feel. I'd forgotten most of its contents within a month, but the title stuck in my imagination like the aura of a half-recollected dream, or a mantra: wild-life-in-a-southern-county. I was living in Hertfordshire then, and the only 'southern' place I'd been to was the beach at Pevensey. But my emotional compass was already set in that direction. South meant the chalk hills at the top of our road, rising towards the summer sun. South meant a view down along a wooded valley, my private heartland, and a thin stream that wound its silver way towards the high Chilterns. In those adolescent years – already an incipient romantic – I would stand at a ritually precise spot at the top of the hill and gaze down that valley in a state of muddled rapture. What I was looking at seemed both wild, numinous, somehow beyond reach and understanding, but also profoundly and anciently English.

No wonder Jefferies chimed with me. *Wild Life* (and note the powerful separation of those two normally conjoined words) is a collection of free-range essays exploring the author's unresolved feelings about the relations between the natural and human worlds. It's set in the very human context of the Wiltshire smallholding where he grew up, but is peopled mostly by non-human species. Jefferies makes the dialectic between these two worlds explicit in his short preface: 'There is a frontier line to civilisation in this country yet, and not far outside its great centres we come quickly even now to the borderland of nature. . . . If we go a few hours journey only, and then step just beyond the highway – where the steam-ploughing engine has left the mark of its wide wheels on the dust – and glance into the hedgerow, the copse, or stream, there are nature's children as unrestrained in their wild, free life as they were in the veritable backwoods of primitive England.'

But the frontier is porous, fluid, debatable. Living things – humans included – pass across it, in both directions. So do ways of perceiving them, which can be coloured both by the civilised, rational mind or the feral imagination. Later in the preface Jefferies writes that 'nature is not cut and dried to hand, nor easily classified, each subject shading gradually into another. In studying the ways, for instance of so common a bird as the starling it cannot be separated from the farmhouse in the thatch of which it often breeds, the rooks with whom it associates, or the friendly sheep upon whose backs it sometimes rides.' This 'shading' of subjects is an exact description of Jefferies' meandering prose-style, and there is, I suspect, an element of rationalisation here of its sometimes chaotic discursiveness. But he was an intuitive ecologist, and this insistence on the connectivity of the natural world is a theme that runs through the book, and justifies his grouping of its contents by habitats. Except that they are not truly natural habitats, but the human landscapes around Coate Farm, near Swindon, where Jefferies was born and lived until he was eighteen: orchard, woodpile, homefield, ash copse, rabbit warren.

But there is sleight of place, and memory, here. Although *Wild Life* is set in Wiltshire, in the present tense, it was written in Surbiton in 1878-9, more than a decade after the encounters it chronicles. Jefferies was then thirty years old and had moved to suburban Surrey to be closer to the London newspaper world. The book is quarried from articles he contributed to the *Pall Mall Gazette*, early examples of what has become an enduring form in British journalism – the 'Country Diary'.

Jefferies' distance from the scenes he is describing (it was also, as we'll see later, a social distancing) helps account for his fascination with borderlands, and for what is a dominant motif in the book. If the thread which runs through Edward Thomas' analogous, echoic *The South Country* (1909) is the pathway, on which he could walk himself up and out of his black moods (Thomas published a biography of Jefferies in the same year), the keynote of Jefferies' *Wild Life* is the hedgerow, in which he can burrow down, away from the messiness of human society. The hedge is the 'frontier line to civilisation'. It is a mark of division in the affairs of humans, but a connective tissue for wildlife. It represents refuge, but also a kind of linear commonland. Jefferies recalls, in a later book, how his father used to point with disgust to 'our Dick poking about in them hedges', and like the poet John Clare, he is most at home – and at his best as a writer – in the hedge-bottom looking out, not on the hilltop taking imperious (or queasily spiritual) views of the landscape below. In one passage he recounts how, peering through a gap in a hedge, he once experienced a kind of optical illusion, in which a hill he knew suddenly appeared vastly higher than it had before. A cloud was resting on its top, and for a while had taken on the exact shape and tone of the hill. With the rest of the range obscured by the hedge, this glimpse through the gap revealed something closer to an alp. The aesthete always lurked, sometimes enlighteningly, sometimes subversively, inside the watchful naturalist.

More literally, hedges were the 'highways' of Jefferies' wild neighbours. Birds and animals passed up and down them between the copses and the farm. One major 'caravan route . . . abuts on the orchard [and] the finches, after spending a little time in the apple and damson trees, fly over the wall and road to [a] second hedge, and follow it down for nearly half a mile to a little enclosed meadow, which, like the orchard, is a specially favourite resort'. It isn't hard to imagine 'our Dick' himself dodging the waves of tits and blossom-haunting goldfinches ('a flood of sunshine falling through a roof of rosy pink') building up a map of their movements and 'resorts', and in the process conjuring the outlines of what sometimes seemed to him the skeleton or ghostly relics of the Wiltshire wildwood.

Jefferies' attention to what he saw is rapt, exact, almost painterly. The look of nature seemed to him as good a guide as any to its meaning and order. He notices the sparkle of ice on the high branches of beech trees in winter, and suggests that this 'proves that water is often present in the atmosphere in large quantities'. His vivid description of a magpie's movements perfectly catches the bird's character, but also begins to explain what it may be up to in its seemingly erratic foragings: 'he walks now to the right, a couple of yards, now to the left in a quick zigzag, so working across the field towards you; then with a long rush he makes a lengthy traverse at the top of his speed, turns and darts away again at right angles, and presently up goes his tail and he throws his head down with a jerk of the whole body as if he would thrust his beak deep into the earth.' He devotes almost two pages to the ripening colours of wheat, noting a moment when it briefly pales during a breeze, 'because the under part of the ear is shown and part of the stalk'. He listens to the heavy buzz of hornets, and peers at them intently enough to know they are the most inoffensive of insects. And he watches a thrush smashing a snail on a sarsen stone: 'about two such blows break the shell, and he then coolly chips the fragments off as you might from an egg'.

There is a kind of hedge-scientist at work behind these observations, thinking by analogy, forging explanations by the application of reason (or at least a particular kind of reason) to acute observation. Jefferies rarely attempts to test his theories methodically, and never quotes the opinions or experiences of any other naturalists. He is an intellectual hermit. This occasionally leads him towards conclusions that today would be regarded as fanciful. He observes the large clutch size and sociable behaviour of long-tailed tits (cousins and unpaired birds often help with the feeding of the young), and concludes that several female birds lay their eggs in one communal nest. A cuckoo lingering close to the nest where she's laid her egg makes him 'doubt the cuckoo's alleged total indifference to her young'. He is also sceptical about cuckoos' host species failing to recognise that the monstrous chick growing in their nest is not one of their own. 'The robin is far too intelligent. Why, then, does he feed the intruder? There is something here approaching to the sentiment of humanity, as we should call it, towards the fellow creature.'

What lies behind these convictions is Jefferies' unusual attitude towards the idea of 'instinct'. He regards this as an inadequate explanation of the behaviour of wild creatures because they so often make mistakes. He tells the story of a party of sand martins attempting to quarry their nest-holes in the mortar of a thick stone wall at Coate Farm. It was a fruitless task, and 'At last, convinced of the impossibility of penetrating the mortar, which was much harder beneath the surface, they went away in a body . . . Instinct, infallible instinct, certainly would not direct these birds to such an unsuitable spot . . . The incident was clearly an experiment, and when they found it unsuccessful, they desisted.' A more conventional scientific explanation would be that it was precisely the martins' instinct for exploring soft stone that led them to the wall. But Jefferies' beguiling and sympathetic interpretation was correct, and far-sighted; he had simply adopted an over-deterministic view of the nature of instinct, seeing it as infallible, or 'blind'. Intelligent experimentation and exploration are now regarded as entirely compatible with broad instinctual drives.

Jefferies' belief in the free-will of other beings, in the maternal cuckoo and the compassionate robin, extended to an insistence that animals felt joy in their lives: 'You may see it in every motion: in the lissom bound of the hare, the playful leap of the rabbit, the song that the lark and the finch must sing'. But, inside *Wild Life* at least, his sympathy with other creatures is patchy and inconsistent. There is a detachment in his prose, which displays plenty of intense curiosity, but little revelation about his own feelings. After a spellbinding and affectionate account of the family life of kingfishers, for instance, he gives, quite casually, as if he had forgotten what he said about joy, instructions about the best way to shoot the birds, especially the youngsters.

The fact is that, at this stage in his life, Jefferies had not yet worked out which side of that 'frontier line' he was on – anchored with civilisation, or on the wing with unrestrained nature. *Wild Life in a Southern County*, his second non-fiction book, is a transitional work, marking the beginnings of a shift away from such simplistic separations of the world. Contrary to the popular image of him as a deep-rooted countryman, Jefferies was a displaced person almost from birth. Aged four, he was despatched from his family's declining farm to live with an aunt in Sydenham. When he was nine he returned home, only to be shunted off to a succession of private schools in Swindon. No wonder he developed into a moody and solitary adolescent. He began reading Rabelais, and spent long days roaming the hill country round Marlborough. When he was sixteen he ran away from home with his cousin, first to France, and then to Liverpool, where he was found by the police and shipped back to Wiltshire. When the smallholding was badly hit by cattle plague in 1865, he left school for good, and started work in Swindon on a new Conservative paper, *The North Wilts Herald*, where he was a jack-of-all-trades reporter and resident short-story writer. At the end of the 1860s he became vaguely ill, left the paper and took a long recuperative holiday in Brussels. He was extravagantly delighted by the women, the fashions, the sophisticated manners, and from letters to his aunt it is clear what he was beginning to think of the philistinism of Wiltshire society.

He returned to Coate Farm in 1871, with no job and no money. His life began to slip into a mould more typical of the anxious, hand-to-mouth existence of the urban freelance, than of a supposed 'son of the soil'. He wrote a play, a memoir of a prospective Member of Parliament, a right-wing pamphlet that ridiculed the advance of popular education. His breakthrough came with a letter to the *Times* in a similar Conservative vein, scorning the habits, intelligence and apathy of the Wiltshire farm labourer. The letter won him sympathy from landowners, and offers of more journalistic work, and for the next few years he wrote copiously on rural affairs for *Fraser's Magazine* and the *Live Stock Journal*.

His increasing journalistic commitments sparked the move to Surbiton, and regular essays for the *Pall Mall Gazette*, in which he reminisces – albeit in an idealistic way – about life back at Coate. His first fully-fledged non-fiction work, *The Gamekeeper at Home*, is made up of pieces written for the *Gazette* between late 1877 and spring '78. It is essentially a tribute to 'the master's' man and an account of the practical business of policing a sporting estate, and maintains the Conservative, deferential tone of his early journalism. The pieces for *Wild Life* appeared in the *Gazette* between 1878 and '79, and though in them is a new sympathy with the farm worker, and the first glimpses of his nature writing potential, flashes of the old right-wing shooting man continue to appear.

Jefferies had only eight years left to live at this point. He developed tuberculosis in 1881, and pain and disenchantment colour the rest of his work. He seems at last to understand the preciousness of life, to be engaged with it, not just as a curious observer but as a fellow being. His beliefs shift radically towards a kind of pantheism, and politically towards libertarian socialism. In his late essays he begins to write of the history, politics, ecology and aesthetics of the land as part of a single complex experience.

These final essays, such as 'Hours of Spring' (1886) and 'Walks in the Wheat-fields' (1887) are his most mature and powerful. But *Wild Life in a Southern County* contains their first buds. To read these essays today is chastening. There is, in the best of them, an electric attentiveness, a noticing, that is hard to aspire to. They are chastening, too, in what they are able to describe – an abundance of bird and insect life that, despite the contemporary passion for slaughter (in which the author played his part), is unimaginable in the modern industrial countryside. The great set-piece of *Wild Life*, 'Rooks returning to roost' is like an epic Victorian narrative painting, full of intense images – the sound of thousands of black wings 'beating the air with slow steady stroke can hardly be compared with anything else in its weird oppressiveness'; full too of a sense of the deep history, the natural 'tradition' of these great nightly migrations. And of one stunning statistic: the 'aerial army's line of march extends over quite five miles in one unbroken corps'. Jefferies did not know this, but he was sending, in a faltering new language, a message in a bottle from a disappearing country.

Richard Mabey
Waveney Valley, 2011

PREFACE

Richard Jefferies

THERE IS A FRONTIER LINE TO CIVILISATION in this country yet, and not far outside its great centres we come quickly even now on the borderland of nature. Modern progress, except where it has exterminated them, has scarcely touched the habits of bird or animal; so almost up to the very houses of the metropolis the nightingale yearly returns to her former haunts. If we go a few hours' journey only, and then step just beyond the highway – where the steam ploughing engine has left the mark of its wide wheels on the dust – and glance into the hedgerow, the copse, or stream, there are nature's children as unrestrained in their wild, free life as they were in the veritable backwoods of primitive England. So, too, in some degree with the tillers of the soil: old manners and customs linger, and there seems an echo of the past in the breadth of their pronunciation.

But a difficulty confronts the explorer who would carry away a note of what he has seen, because nature is not cut and dried to hand, nor easily classified, each subject shading gradually into another. In studying the ways, for instance, of so common a bird as the starling, it cannot be separated from the farmhouse in the thatch of which it often breeds, the rooks with whom it associates, or the friendly sheep upon whose backs it sometimes rides. Since the subjects are so closely connected, it is best, perhaps, to take the places they prefer for the convenience of division, and group them as far as possible in the districts they usually frequent.

The following chapters have, therefore, been so arranged as to correspond in some degree with the contour of the country. Commencing at the highest spot, an ancient entrenchment on the Downs has been chosen as the starting place from whence to explore the uplands. Beneath the hill a spring breaks forth, and, tracing its course downwards, there next come

the village and the hamlet. Still farther the streamlet becomes a broad
brook, flowing through meadows in the midst of which stands a solitary
farmhouse. The house itself, the garden and orchard, are visited by various
birds and animals. In the fields immediately around – in the great hedges
and the copse – are numerous others, and an expedition is made to the
forest. Returning to the farm again as a centre, the rookery remains to be
examined, and the ways and habits of the inhabitants of the hedges. Finally
come the fish and wildfowl of the brook and lake; finishing in the Vale.

R.J.
Coate Farm, 1879

ONE

The Downs

THE MOST COMMANDING DOWN is crowned with the grassy mound and trenches of an ancient earthwork, from whence there is a noble view of hill and plain. The inner slope of the green fosse is inclined at an angle pleasant to recline on, with the head just below the edge, in the summer sunshine. A faint sound as of a sea heard in a dream – a sibilant 'sish, sish', – passes along outside, dying away and coming again as a fresh wave of the wind rushes through the bennets and the dry grass. There is the happy hum of bees – who love the hills – as they speed by laden with their golden harvest, a drowsy warmth, and the delicious odour of wild thyme. Behind the fosse sinks, and the rampart rises high and steep – two butterflies are wheeling in uncertain flight over the summit. It is only necessary to raise the head a little way, and the cool breeze refreshes the cheek – cool at this height while the plains beneath glow under the heat.

Presently a small swift shadow passes across – it is that of a hawk flying low over the hill. He skirts it for some distance, and then shoots out into the air, comes back halfway, and hangs over the fallow below, where there is a small rick. His wings vibrate, striking the air downwards, and only slightly backwards, the tail depressed counteracting the inclination to glide forwards for a while. In a few moments he slips, as it were, from his balance, but brings himself up again in a few yards, turning a curve so as to still hover above the rick. If he espies a tempting morsel he drops like a stone, and alights on a spot almost exactly below him – a power which few birds seem to possess. Most of them approach the ground gradually, the plane of their flight sloping slowly to the earth, and the angle decreasing every moment till it becomes parallel, when they have only to drop their legs, shut their wings, and, as it were, stand upright in the air to find themselves safe on the sward. By that time their original impetus has diminished, and they feel no shock from the cessation of motion. The hawk, on the contrary, seems to descend nearly in a perpendicular line.

The lark does the same, and often from a still greater height, descending so swiftly that by comparison with other birds it looks as if she must be dashed to pieces; but when within a few yards of the ground, the wings are outstretched, and she glides along some distance before alighting. This latter motion makes it difficult to tell where a lark actually does alight. So, too, with snipe: they appear to drop in a corner of the brook, and you feel positive that a certain bunch of rushes is the precise place; but before you get there the snipe is up again under your feet, ten or fifteen yards closer than you supposed, having shot along hidden by the banks, just above the water, out of sight.

Sometimes, after soaring to an unusual elevation, the lark comes down, as it were, in one or two stages: after dropping say fifty feet, the wings are employed, and she shoots forward horizontally some way, which checks the velocity. Repeating this twice or more, she reaches the ground safely. In rising up to sing she often traces a sweeping spiral in the air at first, going round once or twice; after which, seeming to settle on the line she means to ascend, she goes up almost perpendicularly in a series of leaps, as it were – pausing a moment to gather impetus, and then shooting upwards till a mere speck in the sky. When ten or twelve larks are singing at once, all within a narrow radius – a thing that may be often witnessed from these downs in the spring – the charm of their vivacious notes is greater than when one solitary bird alone discourses sweet music which is lost in the blue dome overhead.

At that time they seem to feed only a few minutes consecutively, and then, as if seized with an uncontrollable impulse, rush up into the air to deliver a brief song, descend, and repeat the process for hours. They have a way, too, of rising but six or eight yards above the earth, spreading the wings out and keeping them nearly still, floating slowly forward, all the while uttering one sweet note softly. The sward by the roadside appears to have a special attraction for them; they constantly come over from the arable fields, alight there, and presently return. In the early spring, when lovemaking is in full progress, the cornfields where the young green blades are just showing become the scene of the most amusing rivalry. Far as the eye can see across the ground it seems alive with larks – chasing each other to and fro, round and round, with excited calls, flying close to the surface, continually alighting and springing up again. A gleam of sunshine and a warm south wind bring forth these merry antics. So like in general hue is the lark to the lumps of brown earth that even at a few paces it is difficult to distinguish her. Some seem always to remain in the meadows; but the majority frequent the arable land, and especially the cornfields on the slopes of the downs, where they may be found in such numbers as rival or perhaps exceed those of any other bird.

At first sight starlings seem more numerous; but this arises from their habit of gathering together in such vast flocks, blackening the earth where they alight. But you may walk a whole day across the downs and still find larks everywhere; so that though scattered abroad they probably equal or exceed the starlings, who show so much more. They are by no means timid, being but little disturbed here: you can get near enough to watch every motion, and if they rise it is only to sing. They never seem to know precisely where they are going to alight – as if, indeed, they were nervously particular and must find a clod that pleases them, picking and choosing with the greatest nicety.

Many other birds exhibit a similar trait: instead of perching on the first branch, they hesitate, and daintily decline the bough not quite to their fancy. Blackbirds will cruise along the whole length of a hedge before finding a bush to their liking; they look in several times ere finally deciding. Wood pigeons will make straight for a tree, and slacken speed and show every sign of choosing it, and suddenly, without the slightest cause apparently, go half a mile farther. The partridge which you could vow had dropped just over the hedge has done no such thing; just before touching the ground she has turned at right angles and gone fifty yards down it.

The impression left after watching the motions of birds is that of extreme mobility – a life of perpetual impulse checked only by fear. With one or two exceptions, they do not appear to have the least idea of saving labour by clearing one spot of ground of food before flying farther: they just hastily snatch a morsel and off again; or, in a tree, peer anxiously into every crack and crevice on one bough, and away to another tree a hundred yards distant, leaving fifty boughs behind without examination. Starlings literally race over the earth where they are feeding – jealous of each other lest one should be first, and so they leave a tract all around not so much as looked at. Then, having run a little way, they rise and fly to another part of the field. Each starling seems full of envy and emulation – eager to outstrip his fellow in the race for titbits; and so they all miss much of what they might otherwise find. Their life is so gregarious that it resembles that of men in cities: watching one another with feverish anxiety – pushing and bustling. Larks are much calmer, and always appear placid even in their restlessness, and do not jostle their neighbours.

See – the hawk, after going nearly out of sight, has swept round, and passes again at no great distance; this is a common habit of his kind, to beat round in wide circles. As the breeze strikes him aslant his course he seems to fly for a short time partly on one side, like a skater sliding on the outer edge.

There is a rough grass growing within the enclosure of the earthwork and here and there upon the hills, which the sheep will not eat, so that it remains in matted masses. In this the hares make their forms; and they must, somehow, have a trick of creeping in their places, since many of the grass blades often arch over, and if they sprang into the form heedlessly this could not be the case, as their size and weight would crush it down. When startled by a passer-by the hare – unless there is a dog – goes off in a leisurely fashion, doubtless feeling quite safe in the length of his legs, and after getting a hundred yards or so sits upon his haunches and watches the intruder. Their runs or paths are rather broader than a rabbit's, and straighter – the rabbit does not ramble so far from home; he has his paths across the meadow to the hedge on the other side, but no farther. The hare's track may be traced for a great distance crossing the hills; but while the roads are longer they are much fewer in number. The rabbit makes a perfect network of runs, and seems always to feed from a regular path;

the hare apparently feeds anywhere, without much reference to the runs, which he uses simply to get from one place to another in the most direct line, and also, it may be suspected, as a promenade on which to meet the ladies of his acquaintance by moonlight.

It is amusing to see two of these animals drumming each other; they stand on their hind legs (which are very long) like a dog taught to beg, and strike with the forepads as if boxing, only the blow is delivered downwards instead of from the shoulder. The clatter of their pads may be heard much farther than would be supposed. Round and round they go like a couple waltzing; now one giving ground and then the other, the forelegs striking all the while with marvellous rapidity. Presently they pause – it is to recover breath only; and 'time' being up, to work they go again with renewed energy, dancing round and round, till the observer cannot choose but smile. This trick they will continue till you are weary of watching.

There are holes on the hills, not above a yard deep, and entering the slope horizontally, which are said to be used by the hares more in a playful mood than from any real desire of shelter. Yet they dislike wet; most wild animals do. Birds, on the contrary, find it answers their purpose, grubs and worms abounding at such times. Though the hare is of a wandering disposition he usually returns to the same form, and, if undisturbed, will use it every day for a length of time, at night perhaps being miles away. If hard pressed by the dogs he will leap a broad brook in fine style, but he usually prefers to cross by a bridge. In the evening, as it grows dusk, if you watch from the elevation of the entrenchment, you may see these creatures steal out into the level cornfield below, first one, then two, presently five or six – looming much larger than they really are in the dusk, and seeming to appear upon the scene suddenly. They have a trick of stealing along close to the low mounds which divide arable fields, so that they are unobserved till they turn out into the open ground.

It is not easy to distinguish a hare when crouching in a ploughed field, his colour harmonises so well with the clods; so than an unpractised eye generally fails to note him. An old hand with the gun cannot pass a field without involuntarily glancing along the furrows made by the plough to see if their regular grooves are broken by anything hiding therein. The ploughmen usually take special care with their work near public roads, so that the furrows end on to the base of the highway shall be mathematically straight. They often succeed so well that the furrows look as if traced with a ruler, and exhibit curious effects of vanishing perspective. Along the furrow, just as it is turned, there runs a shimmering light as the eye traces it up. The ploughshare, heavy and drawn with great force, smooths the earth as it cleaves it, giving it for a time a 'face' as it were, the moisture on which reflects the light. If you watch the farmers driving to market, you will see that they glance up the furrows to note the workmanship and look for game; you may tell from a distance if they espy a hare by the check of the rein and the extended hand pointing.

The partridges, too, cower as they hear the noise of wheels or footsteps, but their brown backs, rounded as they stoop, do not deceive the eye that knows full well the irregular shape taken by lumps of earth. Both hares and rabbits may be watched with ease from an elevation, and if you remain quiet will rarely discover your presence while you are above them. They keep a sharp lookout all round, but never think of glancing upwards, unless, of course, some unusual noise attracts attention.

Looking away from the brow of the hill here over the rampart, see, yonder in the narrow hollow a flock is feeding: you can tell even so far off that it is feeding, because the sheep are scattered about, dotted hither and thither over the surface. It is their habit the moment they are driven to run together. Farther away, slowly travelling up a distant down, another flock, packed close, rises towards the ridge, like a thick white mist stealthily ascending the slope.

Just outside the trench, almost within reach, there lies a small white something, half hidden by the grass. It is the skull of a hare, bleached by the winds and the dew and the heat of the summer sun. The skeleton has disappeared, nothing but the bony casing of the head remains, with its dim suggestiveness of life, polished and smooth from the friction of the elements. Holding it in the hand the shadow falls into and darkens the cavities once filled by the wistful eyes which whilom glanced down from the summit here upon the sweet clover fields beneath. Beasts of prey and wandering dogs have carried away the bones of the skeleton, dropping them far apart; the crows and the ants doubtless had their share of the carcass. Perhaps a wound caused by shot that did not immediately check his speed, or wasting disease depriving him of strength to obtain food, brought him low; mayhap an insidious enemy crept on him in his form.

The joy of life of these animals – indeed, of almost all animals and birds in freedom – is very great. You may see it in every motion: in the lissom bound of the hare, the playful leap of the rabbit, the song that the lark and the finch *must* sing; the soft loving coo of the dove in the hawthorn; the blackbird ruffling out his feathers on a rail. The sense of living – the consciousness of seeing and feeling – is manifestly intense in them all, and is in itself an exquisite pleasure. Their appetites seem ever fresh: they rush to the banquet spread by Mother Earth with a gusto that Lucullus never knew in the midst of his artistic gluttony; they drink from the stream with dainty sips as though it were richest wine. Watch the birds in the spring; the pairs dance from bough to bough, and know not how to express their wild happiness. The hare rejoices in the swiftness of his limbs: his nostrils sniff the air, his strong sinews spurn the earth; like an arrow from a bow he shoots up the steep hill that we must clamber slowly, halting halfway to breathe. On outspread wings the swallow floats above, then slants downwards with a rapid swoop, and with the impetus of the motion rises easily. Therefore it is that this skull here, lying so light in the palm of the hand, with the bright sunshine falling on it, and a shadowy darkness in the vacant orbits of the eyes, fills us with sadness. 'As leaves on leaves, so men on men decay'; how much more so with these creatures, whose generations are so short.

If we look closely into the grass here on the slope of the fosse it is animated by a busy throng of insects rushing in hot haste to and fro. They must find it a labour and a toil to make progress through the green forest of grass-blade and moss and heaths and thick thyme bunches, overtopping them as cedars, but cedars all strewn in confusion, crossing and interlacing, with no path through the jungle. Watch this ant travelling patiently onward, and mark the distance traversed by the milestone of a tall bennet.

First up on a dry white stalk of grass lingering from last autumn; then down on to a thistle leaf, round it, and along a bent blade leading beneath into the intricacy and darkness at the roots. Presently, after a prolonged absence, up again on a dead fibre of grass, brown and withered, torn up by the sheep but not eaten: this lies like a bridge across a yawning chasm – the mark or indentation left by the hoof of a horse scrambling up when the turf was wet and soft. Halfway across the weight of the ant overbalances it, slight as that weight is, and down it goes into the cavity: undaunted, after getting clear, the insect begins to climb up the precipitous edge and again plunges into the wood. Coming to a broader leaf, which promises an open space, it is found to be hairy, and therefore impassable except with infinite trouble; so the wayfarer endeavours to pass underneath, but has in the end to work round it. Then a breadth of moss intervenes, which is worse than the vast prickly hedges with which savage kings fence their cities to the explorer, who can get no certain footing on it, but falls through and climbs up again twenty times, and burrows a way somehow in the shady depths below.

Next, a bunch of thyme crosses the path: and here for a lengthened period the ant goes utterly out of sight, lost in the interior, slowly groping round about within, and finally emerging in a glade where your walking stick, carelessly thrown on the ground, bends back the grass and so throws open a lane to the traveller. In a straight line the distance thus painfully traversed may be ten or twelve inches; certainly in getting over it the insect has covered not less than three times as much, probably more – now up, now down, backwards and sideways, searching out a passage.

As this process goes on from morn till night through the long summer's day, some faint idea may be obtained of the journeys thus performed, against difficulties and obstacles before which the task of crossing Africa from sea to sea is a trifle. How, for instance, does the ant manage to keep a tolerably correct course, steering straight despite the turns and labyrinthine involutions of the path? It is never possible to see far in front – half the time not twice its own length; often and often it is necessary to retrace the trail and strike out a fresh one – a step that would confuse most persons even in an English wood with which they were unacquainted.

Yet by some power of observation, perhaps superior in this respect to the abilities of greater creatures, the tiny thing guides its footsteps without faltering down yonder to the nest in the hollow on the bank of the ploughed field. I say by observation, and the exercise of faculties resembling those of the mind, because I have many times tried the supposed unerring instinct of the ant and found it fail: therefore it must possess a power of correcting error which is the prerogative of reason. Ants cannot, under certain conditions, distinguish their own special haunts. Across a garden path I frequented there was the track of innumerable ants; their ceaseless journeyings had worn a visible path leading from the border on one side to the border on the other, where was a tiny hole, into which they each disappeared in turn. Happily the garden was neglected, otherwise the besom of the gardener would have swept away all traces of the highway they had made. Watching the stream of life pouring swiftly along the track, it seemed to me that, like men walking hurriedly in well-known streets, they took no note of marks or bearings, but followed each other unhesitatingly in the groove.

When street pavements are torn up, the human stream disperses and flows out on either side till it discovers by experience the most convenient makeshift passage. What would be the result if this Watling Street of the ants were interrupted? With a fragment of wood I rubbed out three inches of the path worn in the shallow film of soil deposited over the old gravel, smoothing that much down level. Instantly the crowd came to a stop. The foremost ant halted at the edge where the groove now terminated, turned round and had an excited conversation with the next by means of their antennae; a third came up, a fourth and fifth – a crowd collected, in fact. Now, there was no real obstruction – nothing to prevent them from rushing across to the spot where the path recommenced. Why, then, did they pause? Why, presently, begin to explore, right and left, darting to one side and then to the other examining? Was it not because an old and acquired habit was suddenly uprooted? Surely infallible instinct could have carried them across the space of three inches without any trouble of investigation?

In a few seconds one of the exploring parties, making a curve, hit the other end of the path, and the news was quickly spread, for the rest followed almost immediately. Placing a small pebble across the track on another occasion caused almost the same amount of interference with the traffic. Near the hole into which the ants plunged under the border, and on the edge of the bank, so to say, the path they had worn was not visible – the ground was hard and did not take impression: and there, losing the guidance of the groove, they often made mistakes. Instead of hitting the right hole, many of them missed it and entered other holes left by boring worms, and after a short time reappeared to search again, till, finding the cavern, they hastily plunged into it. This was particularly the case when a solitary insect came along. Therefore it would seem that the ant works its way tentatively, and, observing where it fails, tries another place and succeeds.

TWO

A Drought

ONCE NOW AND THEN in the cycle of the years there comes a summer which to the hills is almost like a fever to the blood, wasting and drying up with its heat the green things upon which animal life depends, so that drought and famine go hand in hand. The days go by and grow to weeks, the weeks lengthen to months, and still no rain. The sun pours down his burning rays, which become hotter as the season advances; the sky is blue and beautiful over the hills – beautiful, but pitiless to the bleating flocks beneath. The breeze comes up from the south, bringing with it white clouds sailing at an immense height, with openings between like azure lakes or aerial Mediterraneans landlocked by banks of vapour.

These, if you watch them from the rampart, slowly dissolve, fragments break away from the mass as the edges of the polar glaciers slip off the ice-cliff into the sea, only these are noiseless. The fragment detached grows visibly thinner and more translucent, its margin stretching out in an uneven fringe: the process is almost exactly like the unravelling of a spotless garment, the threads wavering and twisting as they are carried along by the current, diminishing till they fade and are lost in the ocean of blue. This breaking of the clouds is commonly seen in weather that promises to be fine. From the brow here, you may note a solitary cloud just risen above the horizon; it floats slowly towards us; presently it divides into several parts; these, again, fall away in jagged, irregular pieces like flecks of foam. By the time it has reached the zenith these flecks have lengthened out, and shortly afterwards the cloud has entirely melted and is gone. The delicate hue, the contrast of the fleecy white with the deepest azure, the ever-changing form, the light shining through the gauzy texture, the gentle dreamy motion, lend these clouds an exquisite beauty.

After a while the faint breeze increases, but changes in character; it blows steadily, and the 'sish, sish' of the bennets as it rushes through them becomes incessant. A sense of oppression weighs on the chest – in the midst of the wind, on the verge of the hill, you sigh for a breath of air. This is not air: it is simply heat in motion. It is like the simoom of the desert – producing a feeling of intense weariness. Previously the distant ridges of the downs were shaded by a dim haze hovering over them, toning the rolling curves and softening the bolder bluffs. Now they become distinct; each line is drawn clearly and stands out; the definition is like that which occurs before rain, only without the illusion of nearness.

But the hot wind blows and the rain does not come: the sky is open and free from clouds, less blue perhaps, but harder in tint. The nights are bright and clear and warm; you may sit here on the turf till midnight and find no dew, and still feel the languid, enervating influence of the hot blast. This goes in time, and is succeeded by heavy morning mists hanging like a cloak over the hills and filling up the hollows. They roll away as the day advances, and there is the sun bright as ever in the midst of the cloudless sky. The shepherds say the mists carry away the rain; certainly it does not come.

Every now and then promising signs exhibit themselves. A black bank of vapour receives the setting sun, and in the east huge mountainous clouds with beetling precipices and caverns, in which surely the thunder lurks, swell and roll upwards in the hush of the evening. The farmer unrolls his canvas over the new-made hayrick, which is not yet thatched, thinking that a torrent will descend in the night; but no, the morrow is the same. It is a peculiarity of our usually changeable climate that when once the weather has become thoroughly settled either to dry or wet, no signs of alteration are of any value, true as they may be at other times.

So the heat continues and the drought increases. The 'landsprings' breaking out by the sides of the fields have long since disappeared; the true springs run feebly as the stores of water in the interior of the earth gradually grow less. Great cracks open in the clay of the meadows down below in the vale – rifts, wide and deep, into which you may thrust your walking stick to the handle. Up here on the hills the turf grows hard and inelastic; it loses that springy feel under the foot which makes it so pleasant to walk

upon. The grass becomes dull in tint and touches like wire – all the sap
dried from it, and nothing but fibre left. Beneath the chalk is moistureless,
and nothing can grow on it. The by-roads and paths made with the chalk
or rubble glare in the sunlight, and the flints scattered so thickly about the
ploughed fields seem to radiate heat. All things that should look green are
brown and dusty; even the leaves on the elms seem dusty. The wheat only
flourishes, tall and strong – deep tinted yellow here, a ruddy, golden bronze
yonder, with ears full and heavy, rich and glorious to gaze upon. Insects
multiply and replenish the earth after their fashion exceedingly; the spiders
are busy as maybe, not only those that watch from their webs lying in wait,
but those that chase their prey through the grass as dogs do game.

But under the beautiful sky and the glorious sun there rises up a pitiful
cry the livelong day; it is the quavering bleat of the sheep as their strength
slowly ebbs out of them for the lack of food. Green crops and roots fail, the
aftermath in the meadows beneath will not grow, week after week 'keep'
becomes scarcer and more expensive, and there is, in fact, a famine. Of
all animals a starved sheep is the most wretched to contemplate, not only
because of the angularity of outline and the cavernous depressions where
fat and flesh should be, but because the associations of many generations
have given the sheep a peculiar claim upon humanity. They hang entirely on
human help. They watch for the shepherd as though he were their father;
and when he comes he can do no good, so that there is no more painful
spectacle than a fold during a drought upon the hills.

Once upon a time, passing on foot for a distance of some twenty-five
miles across these hills and grassy uplands, I could not help comparing the
scene to what travellers tell us of desert lands and foreign famines. The
whole of that long summer's day, as I hastened southwards, eager for the
beach and the scent of the sea, I passed flocks of dying sheep: in the hollows
by the way their skeletons were here and there to be seen, the gaunt ribs
protruding upwards in the horrible manner that the ribs of dead creatures
do. Crowds of flies buzzed in the air. Upon the hurdles perched the crow,
bold with over-feasting, and hardly turning to look at me, waiting there
till the next lamb should fall and the 'spirit of the beast go downwards'.

Happy England, that experiences these things so seldom, and even then so locally that barely one in ten hears of or sees them!

The cattle of course suffer too; all day long files of water-carts go down into the hollows where the springs burst forth, and at such times half the work of the farm consists in fetching the precious liquid perhaps a mile or more. Even in ordinary summers there is often a difficulty of this kind; and there are some farmhouses whose water for household uses has to be brought fully half a mile. Of recent years more wells have been sunk, but there are still too few for the purpose. The effect of water in determining the settlements of human beings is clearly shown here. You may walk mile after mile on the ridges and pass nothing but a shed; the houses are in the hollows, the 'coombes' or 'bottoms', as they are called, where the springs run. The villages on the downs are generally on a 'bourne', or winter watercourse.

In summer it is a broad winding trench with low green banks, along whose bed you may stroll dry-shod, with the yellow corn on either hand reaching above your head. A few sedges here and there, and that peculiar whitened appearance left when water has passed over vegetation, betoken that once there was a stream. It is like the watercourses and rivers of the East, which are the roads of the traveller till the storm comes, and, lo! in the morning is a rushing flood. Near the village some water is to be seen in the pond which has been deepened out to hold it, and which is, too, kept up here by a spring.

In winter the bourne often has the appearance of a broad brook: you may observe where the current has arranged the small flints washed in from the fields by the rains. As the villages are on the lesser bournes, so the towns are placed on the banks of the rivers these fall into. There may generally be found a row of villages and hamlets on the last slope of the downs, where the hills sink finally away into the plain and vale, so that if any one went along the edge of the hills he would naturally think the district well populated. But if instead of following the edge he penetrated into the interior he would find the precise contrary to be the case. Just at the edge there is water, the 'heads' of the innumerable streams that make the vale so verdant. In the days when wealth consisted chiefly in flocks and herds, men would naturally settle where there were water brooks.

When at last the drought ceases, and the rain does come, it often pours with tropical vehemence; so that the soil of the fields upon the slopes is carried away into the brooks, and the furrows are filled up level with the sand washed out from the clods, the lighter particles of earth floating suspended in the stream, the heavier sand remaining behind. Then, sometimes, as the slow labourer lingers over the ground, with eyes ever bent downwards, he spies a faint glitter, and picks up an antique coin in his horny fingers: coins are generally found after a shower, on the same principle that the gold-seekers wash away the auriferous soil in the 'cradle', and lay bare the yellow atoms. Such coins, too, are sometimes of the same precious metal, ancient and rude. Sometimes the edge of the hoe clinks against a coin, thus at last discovered after so many centuries; yet which for years must have lain so near the surface as to have been turned over and over again by the ploughshare, though unnoticed.

The magnitude of the space enclosed by the earthwork, the height of the rampart and depth of the fosse, show that it was originally intended to be occupied by a large force. With modern artillery, the mitrailleuse, and above all the breech-loading rifle, a comparatively small number of men could hold a commanding position like this: a steep ascent on three sides, and on the fourth a narrow level ridge, easily swept by their fire. But when this entrenchment was thrown up – the chalky earth and flints probably carried up in osier baskets, for they do not seem to have had wheelbarrows in those times – every single yard of rampart required its spear or threatening arrow, so as to present an unbroken rank along the summit. If not, the enemy approaching to close quarters and attacking several places at once would find gaps through which they might pour into the camp. It seems, therefore, evident that these works once sheltered an army; and, looking at their massive character, it is difficult to resist the conclusion that they were not temporary trenches merely, but were permanently garrisoned.

There is another alternative; they may have been a place of refuge for the surrounding population in the nameless wars waged between rival kings. In that case they would, when resorted to, contain a still larger number of persons; women and children and aged men would be included, and to these must be added cattle and sheep. Now, reflecting upon these considerations, and recollecting the remarks previously made upon the lack of water on these hills, the very curious question arises, How did such an army, or such a refugee population with cattle and horses, supply themselves with sufficient water for drinking purposes? The closest examination of the camp itself fails to yield even a suggestion for an answer.

There is not the slightest trace of a well, and it may fairly be questioned whether a well would have been practicable at that date. For this bold brow itself stands high enough; but then, in addition, it is piled on an elevated plateau or tableland, beneath which again is the level at which springs break out. The wells of the district all commence on this tableland or plain. A depression, too, is chosen for the purpose, and their depth is about ninety feet on the average: many are much deeper. But when to this depth the task of digging right down through the hill piled up above the plain is added, the difficulty becomes extreme.

On walking round the entrenchment at the bottom of the fosse, and keeping an eye upon the herbage – the best of all guides – one spot may be noticed where there grows a little of that 'rowetty' grass seen in the damp furrows of the meadows. But there is no sign whatever of a basin or excavation to catch and contain this slight moisture – slight indeed, for the earth is as hard and impenetrable here as elsewhere, and this faint moisture is evidently caused by the rainfall draining down the slope of the rampart. Looking next outside the works for the source of such a supply, a spring will be found in a deep coombe, or bottom, about 800 yards – say, half a mile – from the nearest part of the fosse, reckoning in a straight line. Then, in bringing up water from this spring, which may be supposed to have been done in skins, a double ascent had to be made: first up on to the level plateau, here very narrow, next up the steep down itself. Those only who have had experience of the immense labour of watering cattle on the hills can estimate the work this must have been. An idea is obtained of the value of an elevated position in early warfare, when men for the sake of its advantage were found willing to submit to such toil.

That, however, is not all – foraging parties fetching water must have been liable to be cut off from the main body; there were no cannon then to cover a sortie, and if the enemy were in sufficient force and took possession of the spring, they could compel an engagement, or drive the besieged to surrender rather than endure the tortures of thirst. So that a study of these English hills – widely different as are the conditions of time and place – may throw a strong light upon many an incident of ancient history. There are no traces remaining of any covered way or hollow dyke leading down the slope in the direction of the spring; but some such traces do seem to exhibit themselves in two places – at the rear of the earthwork along the ridge of the hill, and down the steepest and shortest ascent. The first does not come up to the entrenchment, being separated by a wide interval; the latter does, and may possibly have been used as a covered way, though now much obliterated and too shallow for the purpose. The rampart itself is in almost perfect preservation; in one spot the soil has slightly slipped, but form and outline are everywhere distinct.

In endeavouring, however, for a moment to glance back into the unwritten past, and to reconstruct the conditions of some fourteen or fifteen centuries since, it must not be forgotten that the downs may then have presented a different appearance. There is a tradition lingering still that they were in the olden times almost covered with wood. I have tried to fix this tradition – to focus it and give it definite shape; but like a mist visible from a distance yet unseen when you are actually in it, it refuses to be grasped. Still, there it is. The old people say that the king – they have no idea which king – could follow the chase for some forty miles across these hills, through a succession of copses, woods, and straggling covers, forming a great forest. To look now from the top of the rampart over the rolling hills, the idea is difficult to admit at first. They are apparently bare, huge billowy swells of green, with wide hollows cultivated on the lower levels, but open and unenclosed for mile after mile, almost without hedges, and seemingly treeless save for the gnarled and stunted hawthorns – apparently a bare expanse; but more minute acquaintance leads to different conclusions.

Here, to begin with, on the same ridge as the earthwork and not a quarter of a mile distant, is a small clump of wind-harassed trees, growing on the very edge. They are firs and beech, and, though so thoroughly exposed to furious gales, have attained a fair height even in that thin soil. Beech and fir then, can grow here. Away yonder on another ridge is another such a clump, indistinct from the distance: though there is a pleasant breeze blowing and their boughs must sway to it, they appear motionless. With the exception of the poplar, whose tall top as it slowly bends to the blast describes such an arc as to make its motion visible afar, the most violent wind fails to enable the eye to separate the lines of light coming so nearly parallel from the branches of an elm or an oak, even at a comparatively short distance. The tree looks perfectly still, though you know it must be vibrating to the trunk and loosening the earth with the wrench at its anchoring roots.

In more than one of the deep coombes there is a row of elms – out of sight from this post of vantage – whose tops are about level with the plain, where you may stand on the edge and throw a stone into the rook's nest facing you. On a lower spur, which juts out into the valley, is a broad ash wood. Little more than a mile from hence, on the most barren and wildest part of the down, there yet linger some stunted oaks interspersed among the ash copses which to this day are called 'the Chace' and are proved by documentary evidence to stand on the site of an ancient deer forest. A deer forest, too, there is (though seven or eight miles distant, yet on the same range of hills), to this very day tenanted by the antlered stag. Such evidence could be multiplied; but this is enough to establish the fact that for the whole breadth of the hills to have been covered with wood is well within possibility.

I may even go further, and say that, if left to itself, it would in a few generations revert to that condition; for this reason: that when a clump of trees is planted here, experience has shown that it is not so much the wind or the soil which hinders their growth as the attacks of animals wild and tame. Rabbits in cold, frosty weather have a remarkable taste for the bark of the young ash saplings: they nibble it off as clean as if stripped with a knife, of course frequently killing the plant. Cattle – of which a few wander on the hills – are equally destructive to the young green shoots or tops of many trees. Young horses especially will bark almost any smooth-barked tree, not to eat, but as if to relieve their teeth by tearing it off. In the meadows all the young oaks that spring up from dropped acorns out in the grass are invariably torn up by cattle and the still closer-cropping sheep. If the sheep and cattle were removed, and the plough stood still for a century, ash and beech and oak and hawthorn would reassert themselves, and these wide, open downs become again a vast forest, as doubtless they were when the beaver and the marten, the wild boar and the wolf, roamed over the country.

This great earthwork, crowning a ridge from whence a view for many miles could have been obtained over the tops of the primeval trees, must then have had a strangely different strategical position to what it now seemingly occupies in the midst of almost treeless hills. Possibly, too, the powerful effect of so many square miles of vegetation in condensing vapour may have had a distinct influence upon the rainfall, and have rendered water more plentiful than now: a consideration which may help to explain the manner in which these ancient forts were held.

The general deficiency of moisture characteristic of these chalk hills is such that it is said agriculture flourishes best upon them in what is called a 'dropping' summer, when there is a shower every two or three days, the soil absorbing it so quickly. For the grass and hay crops down below in the vale, and for the arable fields there with a stiff heavy soil, on the other hand, a certain amount of dry weather is desirable, else the plough cannot work in its seasons nor the crops ripen or the harvest be garnered in. So that the old saying was that in a drought the vale had to feed the hill, and in a wet year the hill had to feed the vale: which remains true to a considerable extent, so far at least as the cattle are concerned, and was probably true of men and their food also before the importation of corn in such immense quantities placed both alike free from anxiety on that account. This deficiency of moisture being borne in mind, it is a little curious to find ponds of water on the very summit of the down.

Scarcely a quarter of a mile from the earthwork, and on a level with it – close to the clump of firs and beech alluded to previously – there may be seen on this warm summer day a broad, circular, pan-like depression partially filled with water. Being on the very top of the ridge, and only so far sunk as to hold a sufficient quantity, there is little or no watershed to drain into the pond; neither is there a spring or any other apparent source of supply. It would naturally be imagined that in this exposed position, even if filled to the brim by heavy storms of rain, a week of sultry sunshine would evaporate it to the last drop; instead of which, excepting, of course, unusually protracted spells of dry weather such as only come at lengthy intervals, there will always be found some water here; even under the blazing sunshine a shallow pool remains, and in ordinary times the circular basin is half full.

It is of quite modern construction, and, except indirectly, has no bearing upon the water supply of the earthwork, having been made within a few years only for the convenience of the stock kept upon the hill farms. Some special care is taken in puddling the bottom and sides to prevent leakage, and a layer of soot is usually employed to repel boring grubs or worms which would otherwise make their holes through and let the water soak into the thirsty chalk beneath. In wet weather the pond quickly fills; once full, it is afterwards kept up by the condensation of the thick, damp mists, the dew and cloud-like vapours, that even in the early mornings of the hot summer days so frequently cling about the downs. These more than supply the waste from evaporation, so that the basin may be called a dew pond. The mists that hang about the ridges are often almost as laden with moisture as a rain cloud itself. They deposit a thick layer of tiny bead-like drops upon the coat of the wayfarer, which seem to cling after the manner of oil. Though these hills have not the faintest pretensions to be compared with mountains, yet when the rainy clouds hang low they often stroke the higher ridges, which from a distance appear blotted out entirely, and are then receiving a misty shower.

Then there rise up sometimes thick masses of vapour which during the night have gathered over the brooks and water meadows, the marshy places of the vale, and now come borne on the breeze rolling along the slopes; and as these pass over the dew pond, doubtless its colder water condenses that portion which draws down into the depression where it stands. In winter the vapours clinging about the clumps of beech freeze to the boughs, forming, not a rime merely, like that seen in the vale, but a kind of ice-casing, while icicles also depend underneath. Now, if a wind comes sweeping across the hill with sudden blast, these glittering appendages rattle together loudly, and there falls a hail of jagged icy fragments.

There is another such a pond half a mile or more from the earthwork in another direction, but also on a level, making two upon this high and exposed down. Many others are scattered about – they have become more numerous of late years. Several are situated on the lower plateau, which is also dry enough. Toiling over the endless hills in the summer heats, I have often been driven by necessity of thirst to taste a little of the water contained in them, though well knowing the inevitable result. The water has a dead flavour; it is not stagnant in the sense of impurity, but dead, even when quite clear. In a few moments after tasting it, the mouth dries, with a harsh unpleasant feeling, as if some impalpable dusty particles had got into the substance of the tongue. This is caused probably by suspended chalk, of which it tastes; for assuaging thirst, therefore, it is worse than useless in summer: very different is the exquisitely limpid cool liquid which bubbles out in the narrow coombes far below.

The indirect bearing of the phenomena of these dew ponds upon the water supply of the ancient fort is found in the evidence they supply that under different conditions the deposit of moisture here might have been very much larger. The ice formed upon the branches of the beech trees in winter proves that water is often present in the atmosphere in large quantities; all it requires is something to precipitate it. Therefore, if these hills were once clothed with forest, as previously suggested, it appears possible that the primitive inhabitants, after all, may have carried on their agriculture with less difficulty, and have been able to store up water in their camps with greater ease than would be the case at present. This may explain the traces of primeval cultivation to be seen here on the barest, bleakest, and most unpromising hillsides. Such traces may be discovered at intervals all along the slope, on the summit, and near the foot of the down at the rear of the entrenchment.

It is easy to pass almost over them without observing the nearly obliterated marks – the faint lines left on the surface by the implements of men in the days when the first Caesar was yet a living memory. These marks are like some of the little-used paths which traverse the hills: if you look a long way in front you can see them tolerably distinctly, but under your feet they are invisible, the turf being only so slightly worn by wayfarers. So, to find the signs of ancient fields, look for them from a distance as you approach along the slope; then you will see squares and parallelograms dimly defined upon the sward by slightly raised and narrow banks, green with the grass that has grown over them for so many centuries.

They have the appearance sometimes of shallow terraces raised one above the other, rising with the slope of the down. This terrace formation is perhaps occasionally artificial; but in some cases, I think, the natural conformation of the ground has been taken advantage of, having seen terraces where not the faintest trace of cultivation was visible. It is not always easy either to distinguish between the genuine enclosures of ancient days and the trenches left after the decay of comparatively modern fir plantations, which it is usual to surround with a low mound and ditch. Long after the fir trees have died out the green mound remains; but there are rules by which the two, with a little care, may be distinguished.

The ancient field, in the first place, is generally very much smaller; and there are usually three or four or more in close proximity, divided by the faint green ridges, sometimes roughly resembling in ground-plan the squares of a chessboard. The mound that once enclosed a fir plantation is much higher, and would be noticed by the most casual observer. It encircles a wide area, often irregular in shape, oval or circular, and does not present the regular internal divisions of the other – which, indeed, would be unnecessary and out of place in a copse.

It has become the fashion of recent years to break up the sward of the downs, to pare off the turf and burn it, and scatter the ashes over the soil newly turned up by the plough; the idea being mainly to keep more sheep by the aid of turnips and green crops than could be grazed upon the grass. In places it answers – in many others not; after two or three crops the yield sometimes falls to next to nothing. There is a ploughed field here right upon the ridge of the down, close to the ancient earthwork, where in dry summers I have seen ripening oats barely a foot high, and barley equally short. With all the resources of modern agriculture, artificial manure, deeper ploughing, and more complete cleaning, such results do not seem altogether commensurate with the labour bestowed. Of course it is not always so, else the enterprise would be at once abandoned. But when I come to think of the ancient tillage in the terraces upon the barren slopes, I find it difficult to see how, with their rude implements, the men of those times could have procured any sustenance from their soil, unless I suppose the conditions different.

If there was forest all around, to condense the vapours rolling over and deposit a heavy dew or grateful rainfall, then they may have found the stubborn earth more fruitful. Trees and brakes, and thickets, too, would give shelter and protect the rising growth from the bitter winds; while when first tilled the soil itself would be rich from the decay of accumulated leaves, dead boughs, and vegetable matter. So that the terrace gardens may have yielded plentifully then, and were probably surrounded with stockades to protect them from the ravages of the beasts of the forest. Now the very site of the ancient town can scarcely be distinguished: the sheep graze, the lambs gambol gaily over it in the sunshine, and the shepherd dozes hard by on the slope while his dog watches the flock.

A long day of rain is often followed by a moderately fine evening – the clouds breaking up as the sun nears the horizon. It happened one summer evening, after just such a day of continuous showers, that I was in a meadow about two miles distant from the hills. The rain had ceased, and the sky was clear overhead of all but a thin film of cloud, through which the blue was visible in places. But westward there was still a bank of vapour concealing the sinking sun; and eastwards, towards the downs, it was also thick and dark. I walked slowly along with a gun, on the inner side of a great hedge which hid the hills, waiting every now and then behind a projecting bush for a rabbit to come out – a couple being wanted. In heavy rain, such as had lasted all day, they generally remain within their 'buries' – or if one slips out, he usually keeps on the bank, sheltered by stoles and trees, and nibbling a little of the grass that grows there and is comparatively dry. But as evening approaches and the rain ceases, they naturally come forth to break a long fast, and may then be shot.

Some little time passed thus, when, in sauntering along, I came to a gap in the hedge, and glanced through it in the direction of the downs, there partly visible. The idea at once occurred to me that the part of the hills seen through the gap was remarkably high – very much higher and more mountainous than any I had ever visited; and actually, in the abstraction of the moment, half-intent on the rabbits and half perhaps thinking of other things, I resolved to explore that section more thoroughly. Yet, after walking a few yards farther, somehow it seemed singular that the great elevation of this down should never previously have been so apparent. In short, growing curious in the matter, I returned to the gap and looked again.

There was no mistake: there was the down rising up against the sky – a huge dusky mountainous hill, exactly the same in outline as I remembered it, quite familiar, and yet entirely strange. There was the old barn near the foot of the slope; above it the black line of a low hedge and mound; on the summit the same old clump of trees; and lastly, a tall column of black smoke rising upwards, as if from a steam plough at work. It was all just the same, but lifted up into the air – the hill grown into a mountain. A second and longer gaze failed to discover the explanation of the apparition: the eye was completely deceived, and yet the mind was not satisfied. But upon getting up into the gap of the hedge, so as to obtain a better view from the mound, the cause of the illusion was at once visible.

Looking through the gap was like looking through a narrow window, only a short section of the hill being within sight; from the elevation of the mound the whole range of hills could be seen at the same time. Then it became immediately apparent that on either side of this great mountain the continuation of the down right and left remained still at its former level. Upon the central hill a cloud was resting, and had for the time taken its exact shape. The ridge itself was dark, and the dark grey vapour harmonised precisely with its hue; so that the real hill and the cloud merged into each other. Either the barn and clump of trees were reproduced or perhaps enlarged and distorted by the refraction: the seeming column of smoke was a fragment of a blacker colour which chanced to be in a nearly perpendicular position. Even when recognised as such, the illusion was still perfect; nor could the eye separate the hill from the unsubstantial vapour.

As I watched it, the apparent column of smoke bent, and its upper part floated away, enlarging just as smoke, its upward motion overcome by the wind, slowly yields to the current. Soon afterwards the light breeze stretched out one end of the mass of cloud, began to roll up the other, and presently lifted it, revealing the real ridge beneath, which grew momentarily more distinctly defined. Finally the misty bank hung suspended over the down, and slowly sailed eastwards with the wind. Some time afterwards I saw a similar mirage-like enlargement of the down by cloudy vapour resting on it and assuming its contour; but the illusion was not so perfect, because seen from a more open spot, allowing an extended view of the range, and because the cloud was lighter in colour than the hill to which it clung.

These clouds were, of course, passing at a very low elevation above the earth; in rainy weather, although but a few hundred feet high, the ridges are frequently obscured with cloud. The old folk in the vale, whose whole lives have been spent watching and waiting on the weather, say that the hills 'draw' the thunder – that wherever a storm arises it always 'draws' towards them. If it comes from the west it often splits – one storm going along the ridges to the south, and the other passing over detached hills to the northward; so that the basin between is rarely visited by thunder overhead. They have, too, an old superstition – based, apparently, on a text of the Bible – that the thunder always rises originally in the north, though it may reach them from a different direction. For it is their belief also that thunder 'works round'; so that after a heavy storm, say, in the afternoon, when the air has cleared to all appearance, they will tell you that the sunshine and calm are a deception. In a few hours' time, or in the course of the night, the storm will return, having 'worked round': and indeed in that locality this is very often the case. It is to be observed that even a small copse will for a short distance in its rear quite divert the course of a breeze; so that a weathercock placed on the leeside is entirely untrustworthy: if the wind really blows from the south and over the copse, the weathercock will sometimes point in precisely the opposite direction, obeying the undertow of the gale, as it were, drawing backwards.

In summer especially, I fancy, an effect is sometimes produced by a variation in the electrical condition of comparatively small areas, corresponding perhaps with the difference of soil – one becoming more heated than another. Showers are certainly often of a remarkably local character: a walk of half a mile along a road dark from recent rain will frequently bring you to a place where the dust is white and thick as ever, the line of demarcation sharply marked across the highway. In winter rain takes a wider sweep.

From the elevation of the earthwork on the downs – with a view of mile after mile of plain and vale below – it is easy on a showery summer day to observe the narrow limits of the rain. Dusky streamers, like the train of a vast dark robe, slope downwards from the blacker water-carrying cloud above downwards and backwards, the upper cloud travelling faster than the falling drops. Between the hill and the rain yonder intervenes a broad space of several miles, and beyond it again stretches a clear opening to the horizon. The streamers sweep along a narrow strip of country which is drenched with rain, while on either side the sun is shining.

It seems reasonable to imagine that in some way that strip of country acts differently for the time being upon the atmosphere immediately above it. So singularly local are these conditions, sometimes, that one farmer will show you a flourishing crop of roots which was refreshed by a heavy shower just in the nick of time, while his neighbour is loudly complaining that he has had no rain. When the sky is overcast – large masses of cloud, with occasional breaks, passing slowly across it at a considerable elevation without rain – sometimes through these narrow slits long beams of light fall aslant upon the distant fields of the vale. They resemble, only on a greatly lengthened scale, the beams that may be seen in churches of a sunny afternoon, falling from the upper windows on the tiled floor of the chancel, and made visible by motes in the air. So through such slits in the cloudy roof of the sky the rays of the sun shoot downwards, made visible on their passage by the moisture or the motes floating in the atmosphere. They seem to linger in their place as the clouds drift with scarcely perceptible motion; and the labourers say that the sun is sucking up water there.

In the evening of a fine day the mists may be seen from hence as they rise in the meadows far beneath: beginning first over the brooks, a long white winding vapour marking their course, next extending over the moist places and hollows. Higher in the air darker bars of mist, separate and distinct from the white sheet beneath them, perhaps a hundred feet above it, gradually come into sight as they grow thicker and blacker, one here one yonder – long and narrow in shape. These seem to approach more nearly in character to the true cloud than the mist which hardly rises higher than the hedges. The latter will sometimes move or draw across the meadows when there is no apparent wind, not sufficient to sway a leaf, as if in obedience to light and partial currents created by a variation of temperature in different parts of the same field.

Once now and then, looking at this range of hills from a distance of two or three miles on moonless nights, when it has been sufficiently clear to distinguish them, I have noticed that the particular down on which the earthwork is situated shows more distinctly than the others. By day no difference is apparent; but sometimes by night it seems slightly lighter in hue, and stands out more plainly. This may perhaps be due to some unobserved characteristic of the herbage on its slope, or possibly to the chalky subsoil coming there nearer to the surface. The power of reflecting light possessed by the earth, and varied by different soils or by vegetation, is worth observation.

THREE

The Hillside Hedge

A LOW, THICK HAWTHORN HEDGE runs along some distance below the earthwork just at the foot of the steepest part of the hill. It divides the greensward of the down from the ploughed land of the plain, which stretches two or three miles wide, across to another range opposite. A few stunted ash trees grow at intervals among the bushes, which are the favourite resort of finches and birds that feed upon the seeds and insects they find in the cultivated fields. Most of these cornfields being separated only by a shallow trench and a bank bare of underwood, the birds naturally flock to the few hedges they can find. So that, although but low and small in comparison with the copse-like hedges of the vale, the hawthorn here is often alive with birds: chaffinches and sparrows perhaps in the greatest numbers, also yellowhammers.

The colour of the yellowhammer appears brighter in spring and early summer: the bird is aglow with a beautiful and brilliant, yet soft yellow, pleasantly shaded with brown. He perches on the upper boughs of the hawthorn or on a rail, coming up from the corn as if to look around him – for he feeds chiefly on the ground – and uttering two or three short notes. His plumage gives a life and tint to the hedge, contrasting so brightly with the vegetation and with other birds. His song is but a few bars repeated, yet it has a pleasing and soothing effect in the drowsy warmth of summer. Yellowhammers haunt the cornfields principally, though they are not absent from the meadows.

To this hedge the hill magpie comes: some magpies seem to keep almost entirely to the downs, while others range the vale, though there is no apparent difference between them. His peculiar uneven and, so to say, flickering flight marks him at a distance as he jauntily journeys along beside the slope. He visits every fir copse and beech clump on his way, spending some time, too, in and about the hawthorn hedge, which is a favourite spot. Sometimes in the spring, while the corn is yet short and green, if you glance carefully through an opening in the bushes or round the side of the gateway, you may see him busy on the ground. His restless excitable nature betrays itself in every motion: he walks now to the right a couple of yards, now to the left in a quick zigzag, so working across the field towards you; then with a long rush he makes a lengthy traverse at the top of his speed, turns and darts away again at right angles, and presently up goes his tail and he throws his head down with a jerk of the whole body as if he would thrust his beak deep into the earth. This habit of searching the field apparently for some favourite grub is evidence in his favour that he is not so entirely guilty as he has been represented of innocent blood: no bird could be approached in that way. All is done in a jerky, nervous manner. As he turns sideways the white feathers show with a flash above the green corn; another movement, and he looks all black.

It is more difficult to get near the larger birds upon the downs than in the meadows, because of the absence of cover; the hedge here is so low, and the gateway open and bare, without the overhanging oak of the meadows, whose sweeping boughs snatch and retain wisps of the hay from the top of a waggonload as it passes under. The gate itself is dilapidated – perhaps only a rail, or a couple of 'flakes' fastened together with tar-cord: there are no cattle here to require strong fences.

In the young beans yonder the wood pigeons are busy – too busy for the farmer; they have a habit, as they rise and hover about their feeding-places, suddenly shooting up into the air, and as suddenly sinking again to the level of their course, describing a line roughly resembling the outline of a tent if drawn on paper, a cone whose sides droop inward somewhat. They do this too over the ash woods where they breed, or the fir trees; it is not done when they are travelling straight ahead on a journey.

The odour of the bean flower lingering on the air in the early summer is delicious; in autumn when cut the stalk and pods are nearly black, so that the shocks on the side of the hills show at a great distance. The sward, where the slope of the down becomes almost level beside the hedge, is short and sweet and thickly strewn with tiny flowers, to which the bees come, settling on the ground so that as you walk you nearly step on them, and they rise from under the foot with a shrill, angry buzz.

On the other side the plough has left a narrow strip of green running along the hedge: the horses, requiring some space in which to turn at the end of each furrow, could not draw the share any nearer, and on this narrow strip the weeds and wild flowers flourish. The light-sulphur-coloured charlock is scattered everywhere – out among the corn, too, for no cleaning seems capable of eradicating this plant; the seeds will linger in the earth and retain their germinating power for a length of time, till the plough brings them near enough to the surface, when they are sure to shoot up unless the pigeons find them. Here also may be found the wild garlic, which sometimes gets among the wheat and lends an onion-like flavour to the bread. It grows, too, on the edge of the low chalky banks overhanging the narrow waggon-track, whose ruts are deep in the rubble – worn so in winter.

Such places, close to cultivated land yet undisturbed, are the best in which to look for wild flowers; and on the narrow strip beside the hedge and on the crumbling rubble bank of the rough track may be found a greater variety than by searching the broad acres beyond. In the season the large white bell-like flowers of the convolvulus will climb over the hawthorn, and the lesser striped kind will creep along the ground. The pink pimpernel hides on the very verge of the corn, which will be strewn with the beautiful bluebottle flower, whose exquisite hue there is nothing more lovely in our fields. The great scarlet poppy with the black centre, and 'eggs and butter' – curious name for a flower – will, of course, be there: the latter often flourishes on a high elevation, on the very ridges, provided only the plough has been near.

At irregular intervals along the slope there are deep hollows – shallow near the summit, deepening and widening as they sink, till by the hedge at the foot they broaden out into a little valley in themselves. These great green grooves furrow the sides of the downs everywhere, and for that reason it is best to walk either on the ridge or in the plain at the bottom:

if you follow the slope halfway up you are continually descending and ascending the steep sides of these gullies, which adds much to the fatigue. At the mouths of the hollows, close to the hedge, the great flint stones and lumps of chalky rubble rolling down from above one by one in the passage of the years have accumulated: so that the turf there is almost hidden as by a stony cascade.

On the ridge here is a thicket of furze, grown shrub-like and strong, being untouched by woodman's tool; here the rabbits have their 'buries', and be careful how you tread your way among the bushes, for the ground is undermined with innumerable flint pits long abandoned. This is the favourite resort of the chats, who perch on the furze or on the heaps of flints, perpetually iterating their one note, from which their name seems taken. Within the enclosure of the old earthwork itself the flint-diggers have been at work: they occasionally find a few fragments of rusty metal, doubtless relics of ancient weapons; but little worth preserving is ever found there. Such treasures are much more frequently discovered in the cornfields of the plain immediately beneath than here in the camp where one would naturally look for them.

The labourers who pick up these things often put an immensely exaggerated value on them: a worn Roman coin of the commonest kind, of which hundreds are in existence, they imagine to be worth a week's wages, till after refusing its real value from a collector they finally visit a watchmaker whose aqua-fortis test proves the supposed gold to be brass. So, too, with fossils: a man brought me a common echinus, and expected a couple of 'crownds' at least for it; nothing could convince him that, although not often found just in that district, in others they were numerous. The 'crownd' is still the unit, the favourite coin of the labourers, especially the elder folk. They use the word something in the same sense as the dollar, and look with regret upon the gradual disappearance of the broad silver disc with the figure of 'St Gaarge' conquering the dragon.

Everywhere across the hills traces of the old rabbit warrens may be found in the names of places. Warren Farms, Warren Houses, etc., are common; and the term is often added to the names of the villages to distinguish an outlying part of the parish. From the earthwork the sites of four such warrens, now cultivated, can be seen within the radius of as many miles. Rabbits must have swarmed on the downs in the olden times. In the season when the couch and weeds are collected in heaps and burned, the downs – were it not for the silence – might seem the scene of a mighty conflict, the smoke of the battle rolling along the slopes and hanging over the plains, rising up from the hollows in dusky clouds. But the cannon of the shadowy army give forth no thunderous roar. The smouldering fires are not, of course, peculiar to the hills, but the smoke shows so much more at that elevation.

At evening, if you watch the sunset from the top of the rampart, as the red disc sinks to the horizon and the shadows lengthen – the trees below and the old barn throwing their shadows up the slope – the eye is deceived by the position of the light, and the hill seems much higher and steeper, looking down from the summit, than it does at noonday. It is an optical delusion. Here on the western side the grass is still dry – in the deep narrow valleys behind, the sun having long since set over the earthwork and ridge, the dew is already gathering thickly on the sward.

A broad green track runs for many a long, long mile across the downs, now following the ridges, now winding past at the foot of a grassy slope, then stretching away through cornfield and fallow. It is distinct from the waggon-tracks which cross it here and there, for these are local only, and, if traced up, land the wayfarer presently in a maze of fields, or end abruptly in the rickyard of a lone farmhouse. It is distinct from the hard roads of modern construction which also at wide intervals cross its course, dusty and glaringly white in the sunshine. It is not a farm track – you may walk for twenty miles along it over the hills; neither is it the king's highway.

For seven long miles in one direction there is not so much as a wayside tavern; then the traveller finds a little cottage, with a bench under a shady sycamore and a trough for a thirsty horse, situated where three such modern roads (also lonely enough) cross the old green track. Far apart, and far away from its course, hidden among their ricks and trees a few farmsteads stand, and near them perhaps a shepherd's cottage: otherwise it is an utter solitude, a vast desert of hill and plain; silent, too, save for the tinkle of a sheep bell, or, in the autumn, the moaning hum of a distant thrashing machine rising and falling on the wind.

The origin of the track goes back into the dimmest antiquity; there is evidence that it was a military road when the fierce Dane carried fire and slaughter inland, leaving his 'nailed bark' in the creeks of the rivers, and before that when the Saxons pushed up from the sea. The eagles of old Rome, perhaps, were borne along it, and yet earlier the chariots of the Britons may have used it – traces of all have been found; so that for fifteen centuries this track of the primitive peoples has maintained its existence through the strange changes of the times, till now in the season the cumbrous steam-ploughing engines jolt and strain and pant over the uneven turf.

Today, entering the ancient way, eight miles or so from the great earthwork, hitherto the central post of observation, I turn my face once more towards its distant rampart, just visible, showing over the hills a line drawn against the sky. Here, whence I start, is another such a camp, with mound and fosse; beyond the one I have more closely described some four miles is still a third, all connected by the same green track running along the ridges of the downs and entirely independent of the roads of modern days. They form a chain of forts on the edge of the downland overlooking the vale. At starting the track is but just distinguishable from the general sward of the hill: the ruts are overgrown with grass – but the tough 'tussocky' kind, in which the hares hide, avoids the path, and by its edge marks the way. Soon the ground sinks, and then the cornfields approach, extending on either hand – barley, already bending under the weight of the awn, swaying with every gentle breath of air, stronger oats and wheat, broad squares of swede and turnip and dark-green mangold.

Plough and harrow press hard on the ancient track, and yet dare not encroach upon it. With varying width, from twenty to fifty yards, it runs like a green riband through the sea of corn – a width that allows a flock of sheep to travel easily side by side, spread abroad, and snatch a bite as they pass. Dry, shallow trenches full of weeds, and low, narrow mounds, green also, divide it from the arable land; and on these now and then grow storm-stunted hawthorn bushes, gnarled and aged. On the banks the wild thyme grows in great bunches, emitting an exquisite fragrance – luxurious cushions these to rest upon beneath the shade of the hawthorn, listening to the gentle rustle of the wheat as the wind rushes over it. Away yonder the shadows of the clouds come over the ridge, and glide with seeming sudden increase of speed downhill, then along the surface of the corn, darkening it as they pass, with a bright band of light following swiftly behind. It is gone, and the beech copse away there is blackened for a moment as the shadow leaps it. On the smooth bark of those beeches the shepherd lads have cut their names with their great clasp-knives.

Sometimes in the evening, later on, when the wheat is nearly ripe, such a shepherd lad will sit under the trees there; and as you pass along the track comes the mellow note of his wooden whistle, from which poor instrument he draws a sweet sound. There is no tune – no recognisable melody: he plays from his heart and to himself. In a room doubtless it would seem harsh and discordant; but there, the player unseen, his simple notes harmonise with the open plain, the looming hills, the ruddy sunset, as if striving to express the feelings these call forth.

Resting thus on the wild thyme under the hawthorn, partly hidden and quite silent, we may see stealing out from the corn into the fallow: first one, then two, then half a dozen or more young partridge chicks. With them is the anxious mother, watching the sky chiefly, lest a hawk be hovering about; nor will she lead them far from the cover of the wheat. She stretches her neck up to listen and look: then, reassured, walks on, her head nodding as she moves. The little ones crowd after, one darting this way, another that, learning their lesson of life – how and where to find the most suitable food, how to hide from the enemy: imitation of the parent developing hereditary inclinations.

At the slightest unwonted sound or movement she first stretches her neck up for a hurried glance, then, as the labouring folk say, 'quats' – i.e. crouches down – and in a second or two runs swiftly to cover, using every little hollow of the ground skilfully for concealment on the way, like a practised skirmisher. The ants' nests, which are so attractive to partridges, are found in great numbers along the edge of the cornfields, being usually made on ground that is seldom disturbed. The low mounds that border the green track are populous with ants, whose nests are scattered thickly on these banks, as also beside the paths and waggon-tracks that traverse the fields and are not torn up by the plough. Any beaten track such as this old path, however green, is generally free from them on its surface: ants avoid placing their nests where they may be trampled upon. This may often be noticed in gardens: there are nests at and under the edge of the paths, but none where people walk. It is these nests in the banks and mounds which draw the partridges so frequently from the middle of the fields to the edges where they can be seen; they will come even to the banks of frequented roads for the eggs of which they are so fond.

Now that their courting-time is over, the larks do not sing so continuously. Later on, when the ears of wheat are ripe and the reapers are sharpening their sickles, if you walk here, with the corn on either hand, every ten or twenty yards a cloud of sparrows and small birds will rise from it, literally hiding the hawthorn bush on which they settle, so that the green tree looks brown. Wait a little while, and with defiant chirps back they go, disappearing in the wheat.

The sparrows will sometimes flutter at the top of the stalk, hovering for a few moments in one spot, as if trying to perch on the ears; then, grasping one with their claws, they sink with it and bear it to the ground, where they can revel at their leisure. A place where a hailstorm or heavy rain has beat down and levelled the tall corn flat is the favourite spot for these birds; they rise from it in hundreds at a time. But some of the finches are probably searching for the ripe seeds of the weeds that spring up among the corn; they find also a feast of insects.

Leaving now the gnarled hawthorn and the cushion of thyme, I pass a deserted sheep pen, where in the early year the tender lambs were sheltered from the snow and wind. Mile after mile, and still no sign of human life – everywhere silence, solitude. Hill after hill and plain after plain. Presently the turf is succeeded by a hard road – flints ground down into dust by broad waggon wheels bearing huge towering loads of wool or heavy wheat. Just here the old track happens to answer the purposes of modern civilisation. Past this, and again it reverts to turf, leaving now the hills for a mile or two to cross a plain lying between a semicircle of downs; and here at last are hedges of hawthorn and hazel and stunted crab tree.

Round black marks upon the turf, with grey ashes scattered about and half-consumed sticks, show where the gypsies have recently bivouacked, sheltered somewhat at night by the hedges. Near by is an ancient tumulus, on which grows a small yet obviously aged sycamore tree, stunted by wind and storm, and under it the holes of rabbits – drilling their habitations into the tomb of the unknown warrior. In his day, perhaps, the green track wound through a pathless wood long since cleared. Soon the hedges all but disappear, the ground rises once more, nearing the hills; and here the way widens out – first fifty, then a hundred yards across – green sward dotted with furze and some brake fern, and bunches of tough dry grass. Above on the summit is another ancient camp, and below two more turf-grown tumuli, low and shaped like an inverted bowl. Many more have been ploughed down, doubtless, in the course of the years: sometimes still, as the share travels through the soil, there is a sudden jerk, and a scraping sound of iron against stone.

The ploughman eagerly tears away the earth, and moves the stone to find a thin jar, as he thinks – in fact, a funeral urn. But he is imbued with the idea of finding hidden treasure, and breaks the urn in pieces to discover – nothing; it is empty. He will carry the fragments home to the farm, when, after a moment's curiosity, they will be thrown aside with potsherds, and finally used to mend the floor of the cowpen. The track winds away yet farther, over hill after hill; but a summer's day is not long enough to trace it to the end.

In the narrow valley, far below the frowning ramparts of the ancient fort, a beautiful spring breaks forth. Three irregularly circular green spots, brighter in colour than the dry herbage around, mark the outlets of the crevices in the earth through which the clear water finds its way to the surface. Three tiny threads of water, each accompanied by its riband of verdant grasses, meander downwards some few yards, and then unite and form a little stream. Then the water in its channel first becomes visible, glistening in the sun; for at the sources the aquatic grasses bend over, growing thickly, and hide it from view. But pressing these down, and parting them with the hand, you may trace the exact place where it rises, gently oozing forth without a sound.

Lower down, where the streamlet is stronger and has worn a groove – now rushing over a floor of tiny flints, now partly buoyed up and chafing against a smooth round lump of rubble – there is a pleasant murmur audible at a short distance. Still farther from the source, where, grown wider, the shallow water shoots swiftly over a steeper gradient, the undulations of its surface cross each other, plaiting a pattern like four strands interwoven. The resemblance to the pattern of four rushes which the country children delight to plait together as they wander by the brooks is so close as almost to suggest the derivation of the art of weaving rushes, flags, and willows by the hand. The sheep grazing at will in the coombe eat off the herbage too closely to permit many flowers. Where the springs join and irrigate a broader strip there grows a little watercress, and some brooklime, said to be poisonous and occasionally mistaken for the cress; a stray cuckoo flower shows its pale lilac petals in spring, and a few bunches of rushes are scattered round. They do not reach any height or size; they seem dry and sapless, totally unlike the tall green succulent rush of the meadows far below.

A water wagtail comes now and then; sometimes the yellow variety, whose colour in the spring is so bright as to cause the bird to resemble the yellowhammer at the first glance. But besides these the springhead is not much frequented by birds; perhaps the clear water attracts less visible insect life, and, the shore of the stream being hard and dry, there is no moisture where grubs and worms may work their way. Behind the fountain the steep green wall of the coombe rises almost perpendicularly – so steep as not to be climbed without exertion. At the summit are the cornfields of the level plain which here so suddenly sinks without warning. The plough has been drawn along all but on the very edge, and the tall wheat nods at the verge. From thence a strong arm might send a flat round stone skimming across to the other side of the narrow hollow, and its winding course is apparent.

Like a deep groove it cuts a channel up towards the hills, becoming narrower as it approaches; and the sides diminish in height, till at the neck a few rails and a gate can close it, being scarcely broader than a waggon-track. There, at the foot of the down, it ends, overlooked by a barn, the home of innumerable sparrows, whose nests are made under the eaves, everywhere their keen eyes can find an aperture large enough to squeeze into.

Looking down the steep side of the coombe, near the bottom there runs along a projecting ledge, or terrace, like a natural footway. On the opposite side is another corresponding ledge, or green turf-covered terrace; these follow the windings of the valley, decreasing in width as it diminishes, and gradually disappearing. In its broadest part one of them is used as a waggon-track, for which it is admirably adapted, being firm and hard, even smoother of the coombe itself. If it were possible to imagine the waters of a tidal river rising and ebbing up and down this hollow these ledges would form its banks. Their regular shape is certainly remarkable, and they are not confined to this one place. Such steep-sided narrow hollows are found all along the edge of this range of downs, where they slope to the larger valley which stretches out to the horizon. There are at least ten of them in a space of twelve miles, many having similar springs of water and similar terrace-like ledges, more or less perfect. Towards the other extremity of this particular coombe – where it widens before opening on the valley – the spring spreads and occupies a wider channel, beside which there is a strip of osier bed.

When at the fountain-head, and looking down the current, the end of the coombe westwards away from the hills seems to open to the sky; for the ground falls rapidly and the trees hide any trace of human habitation. The silent hills close in the rear capped by the old fort; the silent cornfields come to the very edge above; the silent steep green walls rise on either hand, so near together that the swallows in the blue atmosphere high overhead only come into sight for a second as they shoot swiftly across. In the evening the red sun, enlarged and bulging as if partly flattened, hangs suspended at the very mouth of the trough-like hollow. It is natural in the silence and the solitude for thoughts of the lapse of time to arise – of the endless centuries since, by some slow geological process, this hollow was formed. Fifteen hundred years ago the men of the camp above came hither to draw water; still the spring oozes and flows, and the sun sinks at the western mouth. So too, doubtless, the sun shone into the hollow in the evening cycle upon cycle ere then.

Up the blade of grass here a tiny white-shelled snail has crawled, feeling in its dull, dim way that evening is approaching. The coils of the little shell are exquisitely turned – the workmanship is perfect; the creature within, there can be no question, is equally perfect in its way and finds a joy in the plants on which it feeds. On the ground below, hidden among the fibres near the roots of the grass, lies another tiny shell; but it is empty, the life that once animated it has fled – whither? Presently the falling dew will condense upon it, and at the opening one round drop will stand; after a while to add its mite to the ceaseless flow of the fountain. Could any system of notation ever express the number of these creatures that have existed in the past? If time is measured by the duration of life, reckoned by their short spans, eternity upon eternity has gone by. To me the greatest marvel is the countless, the infinite number of the organisms that have existed, each with its senses and feelings, whose bodies now help to build up the solid crust of the earth. These tiny shells have had millions of ancestors; Nature seems never weary of repeating the same model.

In the osier bed the brook sparrow chatters; there, too, the first pollard willow stands, or rather leans, hollow and aged, across the water. This tree is the outpost of a thousand others that line the banks of the stream for mile after mile yonder down in the valley. How quickly this little fountain grows into a streamlet and then to a considerable brook! – without apparently receiving the waters of any feeders. In the first half-mile it swells sufficiently, if bayed up properly, to drive a mill – as, indeed, many of the springs issuing from these coombes do just below the mouth. In little more than a mile, measuring by its windings, it becomes broad enough to require some effort to leap it, and then deepens into a fair-sized brook.

The rapidity of the increase is accounted for by the fact that every field it passes whose surface inclines towards it is a watershed from which an unseen but considerable drainage takes place. When no brook passes through the fields the water stands and soaks downwards, or evaporates slowly: directly a ditch is opened it fills, and the effect of a stream is not only to collect water till then unseen, but to preserve it from evaporation or disappearance into the subsoil. Probably, if it were possible to start an artificial stream in many places, after a while it would almost keep itself going at times, provided, of course, that the bottom was not porous. Below the mouth of the coombe the water has worn itself a channel six feet deep in the chalk – washing out the flints that now lie at the bottom. Hawthorn bushes bend over it, and great briars uncut since their first shoot was put forth; the elder, too, grows luxuriously, whose white flowers, emitting a rich but sickly odour, the village girls still gather to make elder water to remove freckles. These bushes hide the deep gully in which the current winds its way – so deep that no cattle can get down to drink.

A cottage stands on the very edge a little farther along; the residents do not dip their water from the running stream, but have made a small pool beside it, with which no doubt it communicates, for the pool, or 'dipping place', is ever full of cool, clear, limpid water. The plan is not without its advantages, because the stream itself, though usually clear, is liable to become foul from various causes – such as a flood, when it is white from suspended chalk, or from cattle higher up above the gully coming to slake their thirst and stirring the sandy grit of the bottom. But the little pool long remains clear, because the water from the stream to enter it has to strain itself through the narrow partition of chalky rubble.

So limpid is the current in general, that the idea of seeing trout presently when it shall widen out naturally arises. But before then the soil changes, and clay and loam spoil the clean, sandy, or gravelly bottom trout delight in. In one such stream hard by, however, the experiment of keeping trout has been tried, and with some success: it could be done without a doubt if it were not that after a short course all the streams upon this side of the downs enter the meadows, and immediately run over a mud bottom. With care, a few young fish were maintained in the upper waters, but it was only as an experiment; left to themselves they would speedily disappear, and of course no angling could be thought of.

On the opposite side of the range of hills, where they decline in height somewhat, but still roll on for a great distance, the contrary is the case. The springs that run in that direction pass over a soil that gives a good clear bottom, and gradually assume the character of rivers; narrow and shallow, but clear, sweet and beautiful. There, as you wander over the down, and push your way through one of those extensive nutwoods which grow on the hills, acres and acres of hazel bushes, suddenly you come to the edge of a steep cliff, falling all but perpendicularly, and lo! at the foot is a winding river, bordered by broad green meads dotted with roan-and-white cattle.

Here in the season the angler may be seen skilfully tempting the speckled trout. Across the meads a grove of elm and oak, and the dull red of old houses dimly seen between, and the low dark crenellated tower of a village church. Behind the downs rise again, their slopes in spring a mass of colour – green corn, squares of bright yellow mustard, bright crimson trifolium, and brown fallows.

FOUR

The Village

A SHORT DISTANCE below the cottagers' 'dipping place', the same stream, leaving the deep groove or gully, widens suddenly into a large clear pool, shaded by two tall fir trees and an equally tall poplar. The tops of these trees are nearly level with the plain above the verdant valley in which the stream flows, and, being side by side, the difference in the manner of their growth is strongly contrasted. The branches of the fir gracefully depend, as if weighed downwards by the burden of the heavy deep green fringe they carry – a fringe tipped with bullion in the spring, for the young shoots are of so light a green as to shade into a pale yellow. The branches of the poplar, on the contrary, point upwards – growing nearly vertically; so that the outline of the tree resembles the tip of an immensely exaggerated artist's brush. This formation is ill adapted for nest building, as it affords little or no surface to build on, and so the poplar is but seldom used by birds.

The pool beneath is approached by a broad track – it cannot be called road – trampled into innumerable small holes by the feet of flocks of sheep, driven down here from the hills for periodical washing. At that time the roads are full of sheep day after day, all tending in the same direction; and the little wayside inns, and those of the village which closely adjoins the washpool, find a sudden increase of custom from the shepherds. There is no written law regulating the washing, but custom has fixed it as firmly as an Act of Parliament: each shepherd knows his day, and takes his turn, and no one attempts to interfere with the monopoly of the men who throw the sheep in. The right of wash here is upheld as sternly as if it were a bulwark of the Constitution.

Sometimes a landowner or a farmer, anxious to make improvements, tries to enclose the approach or to utilise the water in fertilising meadows, or in one way or another to introduce an innovation. He thinks perhaps that education, the spread of modern ideas, and the fact that labourers travel nowadays, have weakened the influence of tradition. He finds himself entirely mistaken: the men assemble and throw down the fence, or fill up the new channel that has been dug; and, the general sympathy of the parish being with them and the interest of the sheep farmers behind them to back them up they always carry the day, and old custom rules supreme.

The sheep greatly dislike water. The difficulty is to get them in; after the dip they get out fast enough. Only if driven by a strange dog, and unable to escape on account of a wall or enclosure, will they ever rush into a pond. If a sheep gets into a brook and cannot get out – his narrow feet sink deep into the mud – should he not be speedily relieved he will die, even though his head be above water, from chill and fright. Cattle, on the other hand, love to stand in water on a warm day.

In rubbing together and struggling with the shepherds and their assistants a good deal of wool is torn from the sheep and floats down the current. This is caught by a net stretched across below, and finally comes into the possession of one or two old women of the village, who seem to have a prescriptive right to it, on payment of a small toll for beer-money. These women are also on the lookout during the year for such stray scraps of wool as they can pick up from the bushes beside the roads and lanes much travelled by sheep – also from the tall thistles and briars, where they have got through a gap. This wool is more or less stained by the weather and by particles of dust, but it answers the purpose, which is the manufacture of mops.

The old-fashioned wool mop is still a necessary adjunct of the farmhouse, and especially the dairy, which has to be constantly swilled out and mopped clean. With the ancient spinning wheel they work up the wool thus gathered; and so, even at this late day, in odd nooks and corners, the wheel may now and then be found. The peculiar broad-headed nail which fastens the mop to the stout ashen 'steale', or handle, is also made in the village. I spell 'steale' by conjecture, and according to pronunciation. It is used also of a rake: instead of a rake handle they say rake 'steale'. Having made the mops, the women go round with them to the farmhouses of the district, knowing their regular customers – who prefer to buy of them, not only as a little help to the poor, but because the mops are really very strongly made.

In the meadows of the vale the waters of the same stream irrigate numerous scattered withy beds, pollard willow trees and tall willow-poles growing thickly in the hedges by the brook. The most suitable of these poles are purchased from the farmers by the willow handicraftsmen of the village up here, to be split into thin flexible strips and plaited or woven into various articles. These strips are made into ladies' work baskets and endless knick-knacks. The flexibility of the willow is surprising when reduced to these narrow pieces, scarcely thicker than stout paper. This industry used to keep many hands employed. There were willow looms in the village, and to show their dexterity the weavers sometimes made a shirt of willow – of course only as a curiosity. The development of straw weaving greatly interfered with this business; and now it is followed by a few only, who are chiefly engaged in preparing the raw material to go elsewhere.

From the ash woods on the slopes and the copses of the fields large ash poles are brought, which one or two old men in the place spend their time splitting up for 'flakes' – a 'flake' being a frame of light wood, used after the manner of a hurdle to stop a gap, or pitched in a row to part a field into two. Hurdle-making is another industry; but of late years hurdles have been made on a large scale by master carpenters in the market towns, who employ several men, and undersell the village maker.

The wheelwright is perhaps the busiest man in the place; he not only makes and mends waggon and cart wheels, and the body of these vehicles, but does almost every other kind of carpentering. Sometimes he combines the trade of a builder with it – if he has a little capital – and puts up cottages, barns, sheds, etc., and his yard is strewn with timber. There is generally a mason, who goes about from farm to farm mending walls and pigsties, and all such odd jobs, working for his own hand.

The blacksmith, of course, is there – sometimes more than one – usually with plenty to do; for modern agriculture uses three times as much machinery and ironwork as was formerly the case. At first the blacksmiths did not understand how to mend many of these newfangled machines, but they have learned a good deal, though some of the pieces still have to be replaced from the implement factories if broken. Horses come trooping in to have new shoes put on. Sometimes a village blacksmith acquires a fame for shoeing horses which extends far beyond his forge, and gentlemen residing in the market towns send out their horses to him to be shod. He still uses a ground-ash sapling to hold the short chisel with which he cuts off the glowing iron on the anvil. He keeps bundles of the young, pliant, ground-ash sticks, which twist easily and are peculiarly tough; and, taking one of these, with a few turns of his wrist winds it round the chisel so as to have a long handle. One advantage of the wood is that it gives a little and does not jar when struck.

The tinker, notwithstanding his vagrant habits, is sometimes a man of substance, owning two or more small cottages, built out of his savings by the village mason – the materials perhaps carted for him free by a friendly farmer. When sober and steady, he has a capital trade: his hands are never idle. Milk tins, pots, pans, etc., constantly need mending; he travels from door to door, and may be seen sitting on a stool in the cart house in the farmyard, tinkering on his small portable anvil, with two or three cottagers' children – sturdy, yellow-haired youngsters – intently watching the mystery of the craft.

In despite of machine-sewn boots and their cheapness, the village cobbler is still an institution, and has a considerable number of patrons. The labourers working in the fields need a boot that will keep out the damp, and for that purpose it must be hand-sewn: the cobbler, having lived among them all his life, understands what is wanted better than the artisan of the cities, and knows how to stud the soles with nails and cover toe and heel with plates till the huge boot is literally iron-clad. Even the children wear boots which for their size are equally heavy: many of the working farmers also send theirs to be repaired. The only thing to be remembered in dealing with a village cobbler is, if you want a pair of boots, to order them six months beforehand, or you will be disappointed. The business occupies

him about as long as it takes a shipwright to built a ship.

Under the trees of the lane that connects one part of the village with another stands a wooden post, once stout, now decaying; and opposite it at some distance the remnants of a second. This was a rope walk, but has long since fallen into disuse; the tendency of the age having for a long time been to centralise industry of all kinds. It is true that of late years many manufacturers have found it profitable to remove their workshops from cities into the country, the rent of premises being so much less, water to be got by sinking a well, less rates, and wages a little cheaper. They retain a shop and office in the cities, but have the work done miles away. But even this is distinctly associated with centralisation. The workmen are merely paid human machines; they do not labour for their own hands in their own little shops at home, or as the rope-maker slowly walked backwards here, twisting the hemp under the elms of the lane, afterwards, doubtless, to take the manufactured article himself to market and offer his wares for sale from a stand in the street.

The millwright used to be a busy man here and there in the villages, but the railways take the wheat to the steam mills of cities, and where the watermills yet run, ironwork has supplanted wood. In some few places still the women and girls are employed making gloves of a coarse kind, doing the work at home in their cottages; but the occupation is now chiefly carried on nearer to the great business centres than this. Another extinct trade is that of the bell foundry. One village situated in the hills hard by was formerly celebrated for the church bells cast there, many of which may be found in far distant towers ringing to this day.

Near the edge of the hill, just above the washpool, stands the village church. Old and grey as it is, yet the usage of the pool by the shepherds dates from still earlier days. Like some of the farmhouses farther up among the hills, the tower is built of flints set in cement, which in the passage of time has become almost as hard as the flint itself. The art of chipping flint to a face for the purpose of making lines or patterns in walls used to be carried to great perfection, and even old garden walls may be seen so ornamented.

The tower is large and tall, and the church a great one; or so it appears in comparison with the small population of the place. But it may be that when it was built there were more inhabitants; for some signs remain that here – as in many other such villages – the people have decreased in numbers: the population has shifted elsewhere. An adjacent parish lying just under the downs has now not more than fifty inhabitants; yet in the olden time a church stood there – long since dismantled: the ancient churchyard is an orchard, no one being permitted to dig or plough the ground.

Entering the tower by the narrow nail-studded door, it is not so easy to ascend the winding geometrical stone staircase, in the confined space and the darkness, for the arrow slits are choked with cobwebs and the dust of years. A faint fluttering sound comes from above, as of wings beating the air in a confined space – it is the jackdaws in the belfry; just as the starlings and swallows in the huge old-fashioned chimneys make a similar murmuring noise before they settle. Passing a slit or two – the only means of marking the height which has been reached – and the dull tick of the old clock becomes audible: slow and accompanied with a peculiar grating vibration, as if the frame of the antique works had grown tremulous with age. The dial-plate outside is square, placed at an angle to the perpendicular lines of the tower: the gilding of the hour-marks has long since tarnished and worn away before the storms, and they are now barely distinguishable; and it is difficult to tell the precise time by the solitary pointer, there being no minute-hand.

Past another slit, and the narrow stone steps – you must take care to keep close to the outer wall where they are widest, for they narrow to the central pillar – are scooped out by the passage of feet during the centuries; some, too, are broken, and others are slippery with something that rolls and gives under the foot. It is a number of little sticks and twigs which have fallen down from the jackdaws' nests above: higher up the steps are literally covered with them, so that you have to kick them aside before you can conveniently ascend. These sticks are nearly all of the same size, brown and black from age and the loss of the sap, the bark remaining on. It is surprising how the birds contrive to find so many suitable to their purpose, searching about under the trees; for they do not break them off, but take those that have fallen.

The best place for finding these sticks – and those the rooks use – is where a tree has been felled or a thick hedge cut some months before. In cutting up the smaller branches into faggots the men necessarily frequently step on them, and so break off innumerable twigs too short to be tied up in the bundle. After they have finished faggoting, the women rake up the fragments for their cottage fires; and later on, as the spring advances, the birds come for the remaining twigs, of which great quantities are left. These they pick up from among the grass; and it is noticeable that they like twigs that are dead but not decayed: they do not care for them when green, and reject them when rotten. Have they discovered that green wood shrinks in drying, and that rotten wood is untrustworthy? Rooks, jackdaws, and pigeons find their building materials in this way, where trees or hedges have been cut; yet even then it must require some patience. They use also a great deal of material rearranged from the nests of last year – that is, rooks and jackdaws.

Stepping out at last into the belfry, be careful how you tread; for the flooring is worm-eaten, and here and there planks are loose: keep your foot, if possible, on the beams, which at least are fixed. It is a giddy height to fall from down to the stone pavement below, where the ringers stand. Their ropes are bound round with list or cloth, or some such thing, for a better grasp for the hand. High as it is to this the first floor, if you should attempt to ring one of these bells, and forget to let the rope slip quickly, it will jerk you almost to the ceiling: thus many a man has broken his bones close to the font where he was christened as a child.

Against the wall up here are iron clamps to strengthen the ancient fabric, settling somewhat in its latter days; and, opening the worm-eaten door of the clock case – the key stands in it – you may study the works of the old clock for a full hour, if so it please you; for the clerk is away labouring in the field, and his aged wife, who produced the key of the church and pointed the nearest way across the meadow, has gone to the spring. The ancient building, standing lonely on the hill, is utterly deserted; the creak of the boards underfoot or the grate of the rusty hinge sounds hollow and gloomy. But a streak of sunlight enters from the arrow-slit, a bee comes in through the larger open windows with a low inquiring buzz; there is a chattering of sparrows, the peculiar shrill screech of the swifts, and a 'jack-jack-daw-jack-daw' -ing outside. The sweet scent of clover and of mown grass comes upon the light breeze – mayhap the laughter of haymakers passing through the churchyard underneath to their work, and idling by the way as haymakers can idle.

The name of the maker on the clock shows that it was constructed in a little market town a few miles distant a century ago, before industries were centralised and local life began to lose its individuality. There are sparrows' nests on the wooden case over it, and it is stopped now and then by feathers getting into the works: it matters nothing here; *Festina lente* is the village motto, and time is little regarded. So, if you wish, take a rubbing, with heelball borrowed from the cobbler, of the inscriptions round the rims of the great bells; but be careful even then, for the ringers have left one carelessly tilted, and if the rope should slip, nineteen hundredweight of brazen metal may jam you against the framework.

The ringers are an independent body, rustics though they be – monopolists, not to be lightly ordered about, as many a vicar has found to his cost, having a silent belfry for his pains, and not a man to be got, either, from adjacent villages. It is about as easy to knock this solid tower over with a walking stick as to change village customs. But if towards Christmas you should chance to say to the ringers that such and such a chime seemed rung pleasantly, be certain that you will hear it night after night coming with a throbbing joyfulness through the starlit air – every note of the peal rising clear and distinct at the exact moment of time, as if struck by machinery, yet with a quivering undertone that dwells on the ear after the wave of sound has gone. Then go out and walk in the garden or field, for it is a noble music; remember, too, that it is a music that has echoed from the hills hundreds and hundreds of years. Rude men as they are, these bellringers gratefully respond to the least appreciation of their art.

A few more turns about the spiral staircase, and then step out on the roof. The footstep is deadened by the dull-coloured lead, oxidised from exposure. The tarnished weathercock above revolves so stiffly as to be heedless of the light air – only facing a strong breeze. The irreverent jackdaws, now wheeling round at a safe distance, build in every coign of vantage, no matter how incongruous their intrusion may be – on the wings of an angel, behind the flowing robe of St Peter, or yonder in the niche, grey and lichen-grown, where stood the Virgin Mary before iconoclastic hands dashed her image to the ground. If a gargoyle be broken or choked so that no water comes through it, they will use it, but not otherwise. And they have nests, too, just on the ledge in the thickness of the wall, outside those belfry windows which are partially boarded up. Anywhere, in short, high up and well sheltered, suits the jackdaw.

When nesting time is over, jackdaws seem to leave the church and roost with the rooks; they use the tower much as the rooks do their hereditary group of trees at a distance from the wood they sleep in at other seasons. How came the jackdaw to make its nest on church towers in the first place? The bird has become so associated with churches that it is difficult to separate the two; yet it is certain that the bird preceded the building. Archaeologists tell us that stone buildings of any elevation, whether for religious purposes or defence, were not erected till a comparatively late date in this island. Now, the low huts of primeval peoples would hardly attract the jackdaw. It is the argument of those who believe in immutable and infallible instinct that the habits of birds, etc., are unchangeable: the bee building a cell today exactly as it built one centuries before our era. Have we not here, however, a modification of habit?

The jackdaw could not have originally built in tall stone buildings. Localising the question to this country, may we not almost fix the date when the jackdaw began to use the church, or the battlements of the tower, by marking the time of their first erection? The jackdaw was clever enough, and had reason sufficient to enable him to see how these high, isolated positions suited his peculiar habits; and I am bold enough to think that if the bee could be shown a better mode of building her comb, she would in time come to use it.

In the churchyard, not far from the foot of the tower where the jackdaws are so busy, stands a great square tomb, built of four slabs of stone on edge and a broader one laid on the top. The inscription is barely legible, worn away by the iron-shod heels of generations of ploughboys kicking against it in their rude play, and where they have not chipped it, filled with lichen. The sexton says that this tomb in the olden days was used as the pay-table upon which the poor received their weekly dole. His father told him that he had himself stood there hungry, with the rest – not broken cripples and widows, but strong, hale men, waiting till the loaves were placed upon the broad slab, so that the living were fed literally over the grave of the dead.

The farmers met every now and then in the vestry and arranged how many men each would find work for – or rather partial work – so that the amount of relief might be apportioned. Men coming from a distance, or even from the next parish, were jealously excluded from settling, lest there should be more mouths to feed; if a family, on the other hand, could by any possibility be got rid of, it was exiled. There were more hands than work; now the case is precisely opposite. A grim witness, this old tomb, to a fragment in that history of the people which is now placed above a list of monarchs.

The oldest person in the village was a woman – as is often the case – reputed to be over a hundred: a tidy cottager, well tended, feeble in body, but brisk of tongue. She reckoned her own age by the thatch of the roof. It had been completely new thatched five times since she could recollect. The first time she was a great girl, grown up: her father had it thatched twice afterwards; her husband had it done the fourth time, and the fifth was three years ago. That made about a hundred years altogether.

The straw had lasted better lately, because there were now no great elm trees to drip, drip on it in wet weather. Cottagers are frequently really squatters, building on the waste land beside the highway close to the hedgerow, and consequently under the trees. This dripping on the roof is very bad for thatch. Straw is remarkably durable, even when exposed to the weather, if good in the first place and well laid on. It may be reckoned to last twenty years on an average, perhaps more. Five thatchings, then, made eighty years; add three years since the last thatching; and the old lady supposed she was seventeen or eighteen at the first – i.e. just a century since. But in all likelihood her recollections of the first thatching were confused and uncertain: she was perhaps eight or ten at that time, which would reduce her real age to a little over ninety. A great part of the village had twice been destroyed by fire since she could remember. These fires are, or were, singularly destructive in villages – the flames running from thatch to thatch, and, as they express it, 'wrastling' across the intervening spaces. A pain is said to 'wrastle', or shoot and burn. Such fires are often caused by wood ashes from the hearth thrown on the dust-heap while yet some embers contain sufficient heat to fire straw or rubbish.

The old woman's memories were wholly of a gossipy family history; I have often found that the very aged have not half so much to tell as those of about sixty to seventy years. The next oldest was a man about eighty; all he knew of history was that once on a time some traitor withdrew the flints from the muskets of the English troops, substituting pieces of wood, which, of course, would not ignite the powder, and thus they were beaten. Of date, place, or persons he had no knowledge. He 'minded' a great snowfall when he was a boy, and helping to drag the coaches out and making a firm road for them with hurdles. Once while grubbing a hedge near the road he found five shillings' worth of pennies – the great old 'coppers' – doubtless hidden by a thief. He could not buy so much with one of the new sort of coppers: liked them as King George made best.

An old lady of about seventy, living at the village inn, a very brisk body, seemed quite unable to understand what was meant by history, but could tell me a story if I liked. The story was a rambling narrative of an amour in some foreign country. The lady, to conceal a meeting with her paramour, which took place in the presence of her son, who was an imbecile (or, in her own words, had 'no more sense than God gave him' – a common country expression for a fool), went upstairs and rained raisins on him from the window. The son told the husband what had happened; but, asked to specify the time, could only fix it by, 'When it rained raisins'. This was supposed to be merely a fresh proof of his imbecility, and the lady escaped.

In this imperfect narrative is there not a distorted version of a chapter in the Pentameron? But how did it get into the mind of an illiterate old woman in an out-of-the-way village? Nothing yet of Waterloo, Culloden, Sedgmoor, or the Civil War; but in the end an old man declared that King Charles had once slept in an old house just about to be pulled down. But then 'King Charles' slept, according to local tradition, in most of the old houses in the country. However, I resolved to visit the place.

Tall yew hedges, reaching high overhead, thick and impervious, such as could only be produced in a hundred years of growth and countless clippings, enclosed a green pleasaunce, the grass uncut for many a year, weeds over-running the smooth surface on which the bowls once rolled true to their bias. In the shelter of these hedges, upon the sunny side, you might walk in early spring when the east wind is harshest, without a breath penetrating to chill the blood, warm as within a cloak of sables, enjoying that peculiar genial feeling which is induced by sunshine at that period only, and which is somewhat akin to the sense of convalescence after a weary illness. Thus, sauntering to and fro, your footstep, returning on itself, passed the thrush sitting on her nest calm and confident.

No modern exotic evergreens ever attract our English birds like the true old English trees and shrubs. In the box and yew they love to build; spindly laurels and rhododendrons, with vacant draughty spaces underneath, they detest, avoiding them as much as possible. The common hawthorn hedge round a country garden shall contain three times as many nests, and be visited by five times as many birds, as the foreign evergreens, so costly to rear and so sure to be killed by the first old-fashioned frost.

The thrushes are singularly fond of the yew berry; it is of a sticky substance, sweet and not unpleasant. Holly berries, too, are eaten; and holly hedges, despite their prickly leaves, are favourites with garden birds. It would be possible, I think, to so plan out a garden as to attract almost every feathered creature.

A fine old filbert walk extends far away towards the orchard: the branches meet overhead. In autumn the fruit hangs thick; and what is more exquisite, when gathered from the bough and eaten, as all fruits should be, on the spot? I cannot understand why filbert walks are not planted by our modern capitalists, who make nothing of spending a thousand pounds in forcing-houses. I cannot help thinking that true taste consists in the selection of what is thoroughly characteristic of soil and climate. Those magnificent yew hedges, the filbert walk – all, in fact, are to be levelled to make way for a garish stucco-fronted hunting box, with staring red stables and every modern convenience. The sundial shaft is already heaved up and broken.

The old mansion was used as a grammar school for a great many years, but has been deserted for the last quarter of a century; and melancholy indeed are the silent hollow halls and dormitories. The whitewashed walls are yellow and green from damp, and covered in patches with saltpetre efflorescence; but they still bear the hasty inscriptions scrawled on them by boyish hands – some far back in the eighteenth century. The history of this little kingdom, with its dynasties of tutors and masters, its succeeding generations of joyous youth, might be gathered from these writings on the walls: sketches in burned stick or charcoal of extinct monarchs of the desk; rude doggerel verses; curious jingles of Latin and English words of which every great school has its specimens; dates of day and month when doubtless some daring expedition was carried out; and here and there (originally hidden behind furniture, we may suppose) bitter words of hatred against the injustice of ruling authorities – arbitrary ushers and cruel masters.

The casements, broken and blown in, have permitted all the winds of heaven to wreak their will; and the storms sweeping over from the adjacent downs beat as they choose upon the floor. Within an upper window two swallows' nests have been built against the wall close to the ceiling, and their pleasant twitter greets you as you enter; and so does the whistling of the starlings on the roof. But below the ring of the bricklayer's trowel as he chips a brick has already given them notice to quit.

FIVE

Village Architecture

SOME FEW FARMHOUSES, with cowyards and rickyards attached, are planted in the midst of the village; and these have cottages occupied by the shepherds and carters, or other labourers, who remain at work for the same employer all the year. These cottages are perhaps the best in the place, larger and more commodious, with plenty of space round them, and fair-sized gardens close to the door. The system of hiring for a twelvemonth has been bitterly attacked; but as a matter of fact there can be no doubt that a man with a family is better off when settled in one spot with constant employment, and any number of odd jobs for his wife and children. The cottages not attached to any particular farm – belonging to various small owners – are generally much less convenient; they are huddled together, and the footpaths and rights of way frequently cross, and so lead to endless bickering.

Not the faintest trace of design can be found in the ground plan of the village. All the odd nooks and corners seem to have been preferred for building sites; and even the steep side of the hill is dotted with cottages, with gardens at an angle of forty-five degrees or more, and therefore difficult to work. Here stands a group of elm trees; there half a dozen houses; next a cornfield thrusting a long narrow strip into the centre of the place; more cottages built with the back to the road, and the front door opening just the other way; a small meadow, a well, a deep lane, with banks built up of loose stone to prevent them slipping – only broad enough for one waggon to pass at once – and with cottages high above reached by steps; an open space where three more crooked lanes meet; a turnpike gate, and, of course, a beerhouse hard by it.

Each of these crooked lanes has its group of cottages and its own particular name; but all the lanes and roads passing through the village are known colloquially as 'the street'. There is an individuality, so to say, in these byways, and in the irregular architecture of the houses, which does not exist in the straight rows, each cottage exactly alike, of the modern blocks in the neighbourhood of cities. And the inhabitants correspond with their dwelling in this respect – most of them, especially the elder folk, being characters in their way.

Such old-fashioned cottages are practically built around the chimney; the chimney in the firm nucleus of solid masonry or brickwork about which the low walls of rubble are clustered. When such a cottage is burned down the chimney is nearly always the only thing that remains, and against the chimney it is built up again. Next in importance is the roof, which, rising from very low walls, really encloses half of the inhabitable space.

The one great desire of the cottager's heart – after his garden – is plenty of sheds and outhouses in which to store wood, vegetables, and lumber of all kinds. This trait is quite forgotten as a rule by those who design 'improved' cottages for gentlemen anxious to see the labourers on their estates well lodged; and consequently the new buildings do not give so much satisfaction as might be expected. It is only natural that to a man whose possessions are limited, things like potatoes, logs of wood, chips, odds and ends should assume a value beyond the appreciation of the well-to-do. The point should be borne in mind by those who are endeavouring to give the labouring class better accommodation.

A cottage attached to a farmstead, which has been occupied by a steady man who has worked on the tenancy for the best part of his life, and possibly by his father before him, sometimes contains furniture of a superior kind. This has been purchased piece by piece in the course of years, some representing a little legacy – cottagers who have a trifle of property are very proud of making wills – and some perhaps the last remaining relics of former prosperity. It is not at all uncommon to find men like this, whose forefathers held farms, and even owned them, but fell by degrees in the social scale, till at last their grandchildren work in the fields for wages. An old chair or cabinet which once stood in the farmhouse generations ago is still preserved.

Upon the shelf may be found a few books – a Bible, of course; hardly a cottager who can read is without his Bible – and among the rest an ancient volume of polemical theology, bound in leather; it dates back to the days of the fierce religious controversies which raged in the period which produced Cromwell. There is a rude engraving of the author for frontispiece, title in red letter, a tedious preface, and the text is plentifully bestrewn with Latin and Greek quotations. These add greatly to its value in the cottager's eyes, for he still looks upon a knowledge of Latin as the essential of a 'scholard'. This book has evidently been handed down for many generations as a kind of heirloom, for on the blank leaves may be seen the names of the owners with the inevitable addition of 'his' or 'her' book. It is remarkable that literature of this sort should survive so long.

Even yet not a little of that spirit which led to the formation of so many contending sects in the seventeenth century lingers in the cottage. I have known men who seemed to reproduce in themselves the character of the close-cropped soldiers who prayed and fought by turns with such energy. They still read the Bible in its most literal sense, taking every word as addressed to them individually, and seriously trying to shape their lives in accordance with their convictions.

Such a man, who has been labouring in the hayfield all day, in the evening may be found exhorting a small but attentive congregation in a cottage hard by. Though he can but slowly wade through the book, letter by letter, word by word, he has caught the manner of the ancient writer, and expresses himself in an archaic style not without its effect. Narrow as the view must be which is unassisted by education and its broad sympathies, there is no mistaking the thorough earnestness of the cottage preacher. He believes what he says, and no persuasion, rhetoric, or force could move him one jot. His congregation approve his discourse with groans and various ejaculations. Men of this kind won Cromwell's victories; but today they are mainly conspicuous for upright steadiness and irreproachable moral character, mingled with some surly independence. They are not agitators in the current sense of the term; the local agents of labour associations seem chosen from quite a different class.

Pausing once to listen to such a man, who was preaching in a roadside cottage in a loud and excited manner, I found he was describing, in graphic and rude language, the procession of a martyr of the Inquisition to the stake. His imagination naturally led him to picture the circumstances as corresponding to the landscape of fields with which he had been from youth familiar. The executioners were dragging the victim bound along a footpath across the meadows to the pile which had been prepared for burning him. When they arrived at the first stile they halted, and held an argument with the prisoner, promising him his life and safety if he would recant, but he held to the faith.

Then they set out again, beating and torturing the sufferer along the path, the crowd hissing and reviling. At the next stile a similar scene took place – promise of pardon, and scornful refusal to recant, followed by more torture. Again, at the third and last stile, the victim was finally interrogated, and, still firmly clinging to his belief, was committed to the flames in the centre of the field. Doubtless there was some historic basis for the story; but the preacher made it quite his own by the vigour and life of the local colouring in which he clothed it, speaking of the green grass, the flowers, the innocent sheep, the faggots, and so on, bringing it home to the minds of his audience to whom faggots and grass and sheep were so well known. They worked themselves into a state of intense excitement as the narrative approached its climax, till a continuous moaning formed a deep undertone to the speaker's voice. Such men are not paid, trained, or organised; they labour from goodwill in the cause.

Now and then a woman, too, may be found who lectures in the little cottage room where ten or fifteen, perhaps twenty, are packed almost to suffocation; or she prays aloud, and the rest respond. Sometimes, no doubt, persons of little sincerity practise these things from pure vanity and the ambition of preaching – for there is ambition in cottage life, as elsewhere; but the men and women I speak of are thoroughly in earnest.

Cottagers have their own social creed and customs. In their intercourse, one point which seems to be insisted upon particularly is a previous knowledge or acquaintance. The very people whose morals are known to be none of the strictest – and cottage morality is sometimes very far from severe – will refuse, and especially the women, to admit a strange girl, for instance, to sleep in their house for ample remuneration, even when introduced by really respectable persons. Servant girls in the country where railways even now are few and far between often walk long distances to see mistresses in want of assistance, by appointment. They get tired; perhaps night approaches, and then comes the difficulty of lodging them, if the house happens to be full. Cottagers make the greatest difficulty, unless by some chance it should be discovered that they met the girl's uncle or cousin years ago.

To their friends and neighbours, on the contrary, they are often very kind, and ready to lend a helping hand. If they seldom sit down to a social gathering among themselves, it is because they see each other so constantly during the day, working in the same fields, and perhaps eating their luncheon a dozen together in the same outhouse. A visitor whom they know from the next village is ever welcome to what fare there is. On Sundays the younger men often set out to call on friends at a distance of several miles, remaining with them all day; they carry with them a few lettuces, or apples from the tree in the garden (according to the season), wrapped up in a coloured handkerchief, as a present.

Some of the older shepherds still wear the ancient blue smock frock, crossed with white 'facings' like coarse lace; but the rising generation use the greatcoat of modern make, at which their forefathers would have laughed as utterly useless in the rainstorms that blow across the open hills. Among the elder men, too, may be found a few of the huge umbrellas of a former age, which when spread give as much shelter as a small tent. It is curious that they rarely use an umbrella in the field, even when simply standing about; but if they go a short journey along the highway, then they take it with them. The aged men sling these great umbrellas over the shoulder with a piece of tar cord, just as a soldier slings his musket, and so have both hands free – one to stump along with a stout stick, and the other to carry a flag basket. The stick is always too lengthy to walk with as men use it in cities, carrying it by the knob or handle; it is a staff rather than a stick, the upper end projecting six or eight inches above the hand.

If any labourers deserve to be paid well, it is the shepherds; upon their knowledge and fidelity the principal profit of a whole season depends on so many farms. On the bleak hills in lambing time the greatest care is necessary; and the fold, situated in a hollow if possible, with the down rising on the east or north, is built as it were of straw walls, thick and warm, which the sheep soon make hollow inside, and thus have a cave in which to nestle.

The shepherd has a distinct individuality, and is generally a much more observant man in his own sphere than the ordinary labourer. He knows every single field in the whole parish, what kind of weather best suits its soil, and can tell you without going within sight of a given farm pretty much what condition it will be found in. Knowledge of this character may seem trivial to those whose days are passed indoors; yet it is something to recollect all the endless fields in several square miles of country. As a student remembers for years the type and paper – can see before his eyes the breadth of the margin, the bevel of the binding and hear again the rustle of the stiff leaves of some tall volume which he found in a forgotten corner of a library, and bent over with such delight, heedless of dust and silverfish and the gathered odour of years – so the shepherd recalls *his* books, the fields; for he, in the nature of things, has to linger over them and study every letter: sheep are slow.

When the hedges are grubbed and the grass grows where the hawthorn flowered, still the shepherd can point out to you where the trees stood – here an oak and there an ash. On the hills he has often little to do but ponder deeply, sitting on the turf of the slope, while the sheep graze in the hollow, waiting for hours as they eat their way. Therefore by degrees a habit of observation grows upon him – always in reference to his charge; and if he walks across the parish off duty he still cannot choose but notice how the crops are coming on, and where there is most 'keep'. The shepherd has been the last of all to abandon the old custom of long service. While the labourers are restless, there may still be found not a few instances of shepherds whose whole lives have been spent upon one farm. Thus,

from the habit of observation and the lapse of years, they often become local authorities; and when a dispute of boundaries or water rights or right of way arises, the question is frequently finally decided by the evidence of such a man.

Every now and then a difficulty happens in reference to the old green lanes and bridle-tracks which once crossed the country in every direction, but get fewer in number year by year. Sometimes it is desired to enclose a section of such a track to round off an estate: sometimes a path has grown into a valuable thoroughfare through increase of population; and then the question comes, Who is to repair it? There is little or no documentary evidence to be found – nothing can be traced except through the memories of men; and so they come to the old shepherd, who has been stationary all his life, and remembers the condition of the lane fifty years since. He always liked to drive his sheep along it – first, because it saved the turnpike tolls; secondly, because they could graze on the short herbage and rest under the shade of the thick bushes. Even in the helplessness of his old age he is not without his use at the very last, and his word settles the matter.

In the winter twilight, after a fall of snow, it is difficult to find one's way across the ploughed fields of the open plain, for it melts on the south of every furrow, leaving a white line where it has lodged on the northern side, till the furrows resemble an endless succession of waves of earth tipped with foam-flecks of snow. These are dazzling to the eyes, and there are few hedges or trees visible for guidance. Snow lingers sometimes for weeks on the northern slopes of the downs – where shallow dry dykes, used as landmarks, are filled with it: the dark mass of the hill is streaked like the black hull of a ship with its line of white paint. Field work during what the men call 'the dark days afore Christmas' is necessarily much restricted, and they are driven to find some amusement for the long evenings – such as blowing out candles at the alehouse with muzzle-loader guns for wagers of liquor, the wind of the cap alone being sufficient for the purpose at a short distance.

The children never forget St Thomas's Day, which ancient custom has consecrated to alms, and they wend their way from farmhouse to farmhouse throughout the parish; it is usual to keep to the parish, for some of the old local feeling still remains even in these cosmopolitan times. At Christmas sometimes the children sing carols, not with much success so far as melody goes, but otherwise successfully enough; for recollections of the past soften the hearts of the crustiest.

The young men for weeks previously have been practising for the mumming – a kind of rude drama requiring, it would seem, as much rehearsal beforehand as the plays at famous theatres. They dress in a fantastic manner, with masks and coloured ribbons; anything grotesque answers, for there is little attempt at dressing in character. They stroll round to each farmhouse in the parish, and enact the play in the kitchen or brewhouse; after which the whole company are refreshed with ale, and, receiving a few coins, go on to the next homestead. Mumming, however, has much deteriorated, even in the last fifteen or twenty years. On nights when the players were known to be coming, in addition to the farmer's household and visitors at that season, the cottagers residing near used to assemble, so that there was quite an audience. Now it is a chance whether they come round or not.

A more popular pastime with young men, and perhaps more profitable, is the formation of a brass band. They practise vigorously before Christmas, and sometimes attain considerable proficiency. At the proper season they visit the farms in the evening, and as the houses are far apart, so that only a few can be called at in the hours available after work, it takes them some time to perambulate the parish. So that for two or three weeks about the end of the old and the beginning of the new year, if one chances to be out at night, every now and then comes the unwonted note of a distant trumpet sounding over the fields. The custom has grown frequent of recent years, and these bands collect a good deal of money.

The ringers from the church come too, with their hand bells, and ring pleasant tunes – which, however, on bells are always plaintive – standing on the crisp frozen grass of the green before the window. They are well rewarded, for bells are great favourites with all country people.

What is more pleasant than the jingling of the tiny bells on the harness of the carthorses? You may hear the team coming with a load of straw on the waggon three furlongs distant; then step out to the road, and watch the massive yet shapely creatures pull the heavy weight up the hill, their glossy quarters scarcely straining, but heads held high showing the noble neck, the hoofs planted with sturdy pride of strength, the polished brass of the harness glittering, and the bells merrily jingling! The carter, the thong of his whip nodding over his shoulder, walks by the shaft, his boy ahead by the leader, as proud of his team as the sailor of his craft: even the whip is not to be lightly come by, but is chosen carefully, bound about with rows of brazen rings; neither could you or I knot the whipcord on to his satisfaction.

For there is a certain art even in so small a thing, not to be learned without time and practice; and his pride in whip, harness, and team is surely preferable to the indifference of a stranger, caring for nothing but his money at the end of the week. The modern system – men coming one day and gone the next – leaves no room for the growth of such feelings, and the art and mystery of the craft loses its charm; the harness bells, too, are disappearing; hardly one team in twenty carries them now.

Those who labour in the fields seem to have far fewer holidays than the workers in towns. The latter issue from factory and warehouse at Easter, and rush gladly into the country; at Whitsuntide, too, they enjoy another recess. But the farmer and the labourer work on much the same, the closing of banks and factories in no way interfering with the tilling of the earth or the tending of cattle. In May the ploughboys still remember King Charles, and on what they call 'shick-shack day' search for oak apples and the young leaves of the oak to place with a spray of ash in their hats or buttonholes: the ash spray must have even leaves; an odd number is not correct. To wear these green emblems was thought imperative even within the last twenty years, and scarcely a labourer could be seen without them. The elder men would tell you – as if it had been a great calamity – that they could recollect a year when the spring was so backward that not an oak leaf or oak apple could be found by the most careful search for the purpose. The custom has fallen much into disuse lately: the carters, however, still attach the ash and oak leaves to the heads of their horses on this particular day.

Many village clubs or friendly societies meet in the spring, others in autumn. The day is sometimes fixed by the date of the ancient feast. The club and fête threaten, indeed, to supplant the feast altogether: the friendly society having been taken under the patronage of the higher ranks of residents. Here and there the feast-day (the day on which the church was dedicated) is still remembered, as in this village, where the elder farmers invite their friends and provide liberally for the occasion. Some of the gypsies still come with their stalls, and a little crowd assembles in the evening; but the glory of the true feast has departed.

The elder men, nevertheless, yet reckon by the feast-day; it is a fixed point in their calendar, which they construct every year, of local events. Such and such a fair is calculated to fall so many days after the first full moon in a particular month; and another fair falls so long after that. An old man will thus tell you the dates of every fair and feast in all the villages and little towns ten or fifteen miles round about. He quite ignores the modern system of reckoning time, going by the ancient ecclesiastical calendar and the moon. How deeply the ancient method must have impressed itself into the life of these people to still remain a kind of instinct at this late day!

The feasts are in some cases identified with certain well-recognised events in the calendar of nature; such as the ripening of cherries. It may be noticed that these, chancing thus to correspond pretty accurately on the average with the state of fruit, are kept up more vigorously than those which have no such aid to the memory. The Lady Day fair and Michaelmas fair at the adjacent market town are the two best recognised holidays of the year. The fair is sometimes called 'the mop', and stalwart girls will walk eight or nine miles rather than miss it. Maidservants in farmhouses always bargain for a holiday on fair day. These two main fairs are the Bank Holidays of rural life. It is curious to observe that the developments of the age, railroads and manufactories, have not touched the traditional prestige of these gatherings.

For instance, you may find a town which, by the incidence of the railroad and the springing up of great industries, has shot far ahead of the other sleepy little places; its population may treble itself, its trade be ten times as large, its attractions, one would imagine, incalculably greater. Nothing of the kind: its annual fair is not nearly so important an event to the village mind as that of an old-world slumberous place removed from the current of civilisation. This place, which is perhaps eight or nine miles by road, with no facilities of communication, has from time immemorial had a reputation for its fair. There, accordingly, the scattered rural population wends, making no account of distance and very little of weather; it is a country maxim that it always rains on fair day, and mostly thunders. There they assemble and enjoy themselves in the old-fashioned way, which consists of standing in the streets, buying 'fairings' for the girls, shooting for nuts, visiting all the shows, and so on.

To push one's way through such a crowd is no simple matter; the countryman does not mean to be rude, but he has not the faintest conception that politeness demands a little yielding. He has to be shoved, and makes no objection. A city crowd is to a certain extent mobile – each recognises that he must give way. A country crowd stands stock-still.

The thumping of drums, the blaring of trumpets, the tootling of pan pipes in front of the shows, fill the air with a din which may be heard miles away, and seem to give the crowd intense pleasure – far more than the crack band of the Coldstream Guards could impart. Nor are they ever weary of gazing at the 'pelican of the wilderness' as the showman describes it – a mournful bird with draggled feathers standing by the entrance, a traditional part of his stock-in-trade. One attraction – perhaps the strongest – may be found in the fact that all the countryside is sure to be there. Each labourer or labouring woman will meet acquaintances from distant villages they have not seen or heard of for months. The rural gossip of half a county will be exchanged.

In the autumn after the harvest the gleaning is still an important time to the cottager, though nothing like it used to be. Reaping by machinery has made rapid inroads, and there is not nearly so much left behind as in former days. Yet half the women and children of the place go out and glean, but very few now bake at home; they have their bread from the baker, who comes round in the smallest hamlets. Possibly they had a more wholesome article in the olden time, when the wheat from their gleanings was ground at the village mill, and the flour made into bread at home. But the cunning of the mechanician has invaded the ancient customs; the very sheaves are now to be bound with wire by the same machine that reaps the corn. The next generation of country folk will hardly be able to understand the story of Ruth.

SIX

The Hamlet

IN MOST LARGE RURAL PARISHES there is at least one small hamlet a mile or two distant from the main village. A few houses and cottages stand loosely scattered about the fields, no two of them together; so separated by hedges, meadows, and copses as hardly to be called even a hamlet. The communication with the village is maintained by a long, winding narrow lane; but foot passengers follow a shorter path across the fields, which in winter is sure to be ankle deep in mud, by the gateways and stiles. The lane, at the same time, is crossed by a torrent, which may spread out to thirty yards wide in the hollow, shallow at the edges, but swift and deep in the middle.

If you wait a couple of hours it will subside, as the farmers lower down the brook pull up the hatches to let the flood pass. If you are in a hurry, you must climb up into the double-mound beside the lane, and force your way along it between thorns and stoles till you reach the channel through which the current is rushing. Across that an old tree trunk will probably lie, and by grasping a bough as a handrail it is possible to get over. But either way, by lane or footpath, you are sure to get what the country folk call 'watchet', i.e. wet. So that in winter the hamlet is practically isolated; for even in moderately good weather the lane is an inch or two deep in finely puddled adhesive mud. It is so shaded by elms and thick hedges that the dirt requires a length of time to dry, while the passage of hundreds of sheep tread and puddle it as only sheep can.

In summer the place is lovely; but then the inhabitants are one and all busy in the fields, and have little time for social intercourse or for travel into the next parish. It is ten to one if you knock at a cottage door you will find it locked – if, indeed, you get so far as that, a padlock being often on the garden gate. Being so isolated, and apart from the current of modern life and manners, the hamlet folk retain something of the old-fashioned way of thinking. They do not believe their own superstitions with the implicit credence of yore, but they have not yet forgotten them. I have known women, for instance, who seriously asserted that such-and-such an aged person possessed a magic book which contained spells, and enabled her to foresee some kinds of coming events. The influence of the moon, so firm an article of faith among their forefathers, is not altogether overlooked; and they watch for the new moon carefully. If the crescent slopes, it will be wet weather. But if the horns of the crescent touch, or nearly, a vertical line, if it stands upright, then it will be fine. Something, too, must be allowed for the degree of sharpness of definition of the crescent, which reveals the state of the atmosphere. And the cottage astrologer has a whole table of the quarters, aspects, and so on, and lays much stress upon the day and hour of the change: indeed, it is a very complicated business to understand the moon.

The belief in the power of certain persons to 'rule the planets' is profound; so profound that neither ridicule, argument, nor authority will shake it in the minds of the hamlet girls, and it abides with them even when they are placed amidst the disenchanting realities of town life. When 'in service', they buy dream books and consult fortune tellers. The gypsies, in passing through the country, choose the byways and lanes; they thus avoid the tolls, have a chance of poaching, and find waste places to camp in, though possibly something of the true nomadic instinct may urge them to leave the beaten tracks and wander over lonely regions. They camp near the hamlet as they travel to and from the great sheep fairs which are held upon the hills, and perhaps stay a few days; and by them, to some extent, the belief in astrology and palmistry is strengthened.

The carters, who have to spend some considerable time every day with their horses in the stable, still retain a large repertory of legendary ghost lore. They know the exact spot in the lane where, at a certain hour of the night, the white spectre of a headless horse, rushing past with incredible swiftness and without the sound of a hoof, brushes the very coat of the traveller, and immediately disappears in the darkness. Another lane is haunted by a white woman, whose spectre crosses it in front of the spectator and then appears behind him. If he turns his head or looks on one side in order to escape the sight of the apparition, it instantly crosses to that side. Indeed, no matter in which direction he glances, the flickering figure floats before him, till, making a run for it, he passes beyond the limits of the haunted ground.

Near by the hollow, where the stream crosses the lane, is another spirit, but of an indefinite kind, that does not seem to take shape, but causes those who go past at the time when it has power to feel a mortal horror. A black dog may be seen in at least two different places: the wayfarer is suddenly surprised to find a gigantic animal of the deepest jet trotting by his side, or he sees a dark shadow detach itself from the bushes and take the form of a dog. The black dog has perhaps more vitality, and survives in more localities, than all the apparitions that in the olden times were sworn to by persons of the highest veracity. They may still be heard of in many a nook and corner. I have known people of the present day who were positive that there really was something weird in the places where the dog was said to appear.

It is supposed that horses are peculiarly liable to take fright and run away, to shy, or stumble and break their knees, at a certain spot in the road. They go very well till just on passing the fatal spot a sudden fear seizes them as if they could see something invisible to men; sometimes they bolt headlong, sometimes stand stock-still and shiver, or throw the rider by a rapid side-movement. In the daytime – for this supernatural effect is felt in broad day as well as at night – the horse more frequently falls or stumbles, as if checked by an invisible force in the midst of his career. This, too, is a living superstition, and some persons will recount a whole string of accidents that have happened within a few yards; till at last, such is the force of iteration, the most incredulous admit it to be a series of remarkable coincidences. These last two, the black dog and the dangerous place in the road, are believed in by people of a much higher grade than carters. Altogether, the vitality of superstition in the country is very much greater than is commonly suspected. It is now confined to the inner life of the people: no one talks of such things openly, but only to their friends, and thus a stranger might remark on the total extinction of the belief in the supernatural. But much really remains.

The carters have a story about horses which had spent the night in a meadow being found the next morning in a state of exhaustion, as if they had been ridden furiously during the hours of darkness. They were totally unfit for work next day. Instances are even given where men have hidden in a tree with a gun, and when the horses began to gallop fired at something indistinct sitting on their haunches, which something at once disappeared, and the excitement ceased. But these things are said to have happened a long time ago.

So, also, there is a memory of a man digging stone in a quarry and distinctly hearing the strokes of a pick beneath him. When he wheeled his barrow the subterranean quarryman wheeled his, and shortly after he had shot the stones out there came a rumbling from below as if the other barrow had been emptied. The very quarry is pointed out where this extraordinary phenomenon took place. It is curious how a story of this kind, something like which is, I think, told of the Hartz Mountains, should have got localised in a limestone quarry so far apart in distance and character. How well I remember the ancient labourer who told me this legend as a boy! It is easy to philosophise on it now, and speculate upon the genesis of the tale, which may have originated in a cavernous hollow resounding to the tools; but then it was a reality, and I recollect always giving a wide berth to that quarry at night. As the old man told it, it was indeed hardly a legend; for he could disclose every detail, and what has here occupied a few sentences took him the best part of an hour to relate.

Now and then the western clouds after the sunset assume a shape resembling that of a vast extended wing, as of a gigantic bird in full flight – the extreme tip nearly reaching the zenith, the body of the bird just below the horizon. The resemblance is sometimes so perfect that the layers of feathers are traceable by an imaginative eye. This, the old folk say, is the wing of the Archangel Michael, and it bodes no good to the evil ones among the nations, for he is on his way to execute a dread command.

Herbs are still believed in implicitly by some. Not long since I met a labourer whom I had known previously, and now found deeply depressed because of the death of a son. The poor fellow had had every attention; the clergyman had exerted himself, and wine and nourishing luxuries had not been spared, nor the best of medical advice. That he admitted, but still regretted one thing. There was a herb, which grew beside rivers, and was known to but a few, that was a certain cure for the kind of wasting disease which had baffled educated skill. There was an old man living somewhere by a river fifty miles away, who possessed the secret of this herb, and by it had accomplished marvellous cures. He had heard of him, but could not by any inquiry ascertain his exact whereabouts, and so his child died. Everything possible had been done, but still he regretted that this herb had not been applied.

Nothing is done right now, according to the old men of the hamlet; even the hayricks are built badly and 'scamped'. The rickmaker used to be an important person, generally a veteran, who had to be conciliated with an extra drop of good liquor before he could be got to set to work in earnest. Then he spread the hay here, and worked it in there, and had it trodden down at the edge, and then in the middle, and, like the centurion, sent men hither and thither. His rick, when complete, did not rise perpendicularly, but each face or square side sloped a little outwards – including the ends – a method that certainly does give the rick a very shapely look.

But now the newfangled 'elevator' carries up the hay by machinery from the waggon to the top, and two ricks are run up while they would formerly have just been carefully laying the foundation for one of faggots to keep off the damp. The poles put up to support the rick-cloth interfere with the mathematically correct outward slope at the ends, upon which the old fellow prided himself; so they are carried up straight like the end wall of a cottage, and are a constant source of contempt to the ancient invalid. However, he consoles himself with the reflection that most of the men employed with the 'elevator' will ultimately go to a very unpleasant place, since they are continuously swearing at the horse that works it, to make him go round the faster.

After an old cart or waggon has done its work and is broken up, the wooden axletree, which is very solid, is frequently used for the top bar of a stile. It answers very well, and, being of seasoned wood that has received a good many coats of red paint, will last a long time. The life of a waggon is not unlike that of a ship. On the cradle it is the pride of the craftsman who builds it, and who is careful to reproduce the exact lines which he learned from his master as an apprentice, and which have been handed down these hundred years and more. The builders of the Chinese junks are said never to saw a piece of timber into the shape required, nor to bend it by softening the fibres by hot steam, but always use a beam that has grown crooked naturally. This plan gives great strength, but it must take years to accumulate the necessary curved trees. The waggon-builder, in like manner, has a whole yardful of timber selected for much the same reason – because it naturally curves in the way he desires, or is specially fitted for his purpose.

For, like the ship, the true old-fashioned waggon is full of curves, and there is scarcely a straight piece of wood about it. Nothing is angular or square; and each piece of timber, too, is carved in some degree, bevelled at the edges, the sharp outline relieved in one way or another, and the whole structure like a ship, seeming buoyant, and floating easily on the wheels. Then the painting takes several weeks, and after that the lettering of the name; and when at last completed it is placed outside by the road, so that every farmer and labourer who goes by may pause and admire. In about twelve months, if the builder be expeditious (for him), the new vessel may reach her port under the open shed at the farm, and then her life of voyages begins.

Her cargoes are hay and wheat and huge mountainous loads of straw, and occasionally hurdles for the shepherd. Nor are her voyages confined to the narrow seas of the fields adjoining home; now and then she goes on adventurous expeditions to distant market towns, carrying mayhap a cargo of oak bark, stripped from fallen trees, to the tan yard. Then she is well victualled for the voyage, and her course mapped out on the chart in order to avoid the Scylla of steep hills and stony ways and the Charybdis

of tollgates, beside being duly cautioned against the sirens that chant so sweetly from the taps of the roadside inns. Or she sails down to the faraway railway station after coal – possibly two or more vessels in the same convoy – if the steam plough be at work and requires the constant services of these tenders.

She has her own special crew – her captain the carter – and for forecastle men a lad or two, and often a couple of able-bodied seamen in the shape of labourers, to help to load up.

When on the more distant voyages to unknown shores, she takes a supercargo – the farmer's son – to check the bills of lading; for on those strange coasts who knows what treachery there may be brewing? There are arms aboard, in the form of forks or prongs; and commonly one or more passengers go out in her – women with vast bundles and children – not to mention the merchandise of sugars and of teas from Cathay, which are shipped for delivery at half the cottages and farmsteads *en route* homewards. Wherefore, you see, the captain needs to be a sober and godly man, having all these and manifold other responsibilities upon his mind.

Besides which he has to make a report upon the state of the crops on every farm he passes, and what everybody is doing, and if they have begun reaping; also to hail every vessel he passes outward or homeward bound, and enter her answers in his log, and to keep his weather-eye open and a sharp watch to windward, lest storms should arise and awake the deep, and if the gale increases to batten down his hatches and make all snug with the tarpaulin. He must bear in mind the longitude of these ports where there are docks, lest his team should cast a shoe or any of the running rigging want splicing, or the hull spring a leak – for the blacksmith's forges are often leagues apart, and he may lose his certificate if he strands his ship or founders on the open ocean of the downs. Sometimes, if the currents run unexpectedly strong, and he is deeply laden, he has to borrow or hire a tug from the nearest farm, getting an extra horse to pull up the hill.

When he reaches harbour, and has leave ashore, a jollier seaman never cracked a whip. Perhaps the happiest time with the ploughboys is when they are out with the waggon, having a little change, no harder work than walking, sips at the 'pots' handed to the captain by his mates, and nothing to think about. Nor was there ever a more popular song in the country than:

> We'll jump into the waggon,
> And we'll all take a ride!

Though in winter, when the horses' shoes have to be roughed for the frost, or, worse, when the wheels sink deep into the spongy turf, and rain and sleet and snow make the decks slippery, it is not quite so jolly. Yet even then, so strong is the love of motion, a run with the waggon is preferred to stationary work.

The captain, when bound on a voyage, generally slips his cable or weighs anchor with the rising sun. His crew are first-rate helmsmen: and to see them sweep into the rickyard through the narrow gateway, with a heavy deck cargo piled to the skies, all sail set, a stiff breeze, and the timbers creaking is a glorious sight! Not a scrape against the jetty, though 'touch and go' is the sign of a good pilot. His greatest trouble is when his cargo shifts out of sight of land: sometimes the vessel turns on her beam-ends with a too ponderous and ill-built load of straw, and then the wreck lies right in the fairway of all the ships coming up the channel. To load a waggon successfully is indeed a work of art: on the hills, where the waggons have to run 'sidelong' to pick up the crops, one side higher than the other, no one but an experienced hand can make the stuff stay on. Then there is often a tremendous bumping and scraping of the keel on the rocks of the newly-mended roads, and the nasty chopping seas of the deep ruts, besides the long regular Atlantic swells to the furrows and 'lands'. So that the cargo had need to be firmly placed in the hold.

Every now and then she goes into dock and gets a new streak of paint and a thorough overhauling. The running rigging of the harness has to be polished and kept in good condition, and the crew are rarely idle if the captain knows his business. You should never let your 'fo' castle' hands loll about; the proverb about the — and the idle hands is notoriously true aboard ship, and in the stables.

How many a man's life has centred about the waggon! As a child he rides in it as a treat to the hayfield with his father; as a lad he walks beside the leader, and gets his first ideas of the great world when they visit the market town. As a man he takes command and pilots the ship for many a long, long year. When he marries, the waggon, lent for his own use, brings home his furniture. After a while his own children go for a ride in it, and play in it when stationary in the shed. In the painful ending the waggon carries the weak-kneed old man in pity to and from the old town for his weekly store of goods, or mayhap for his weekly dole of that staff of life his aged teeth can hardly grind. And many a plain coffin has the old waggon carried to the distant churchyard on the side of the hill. It is a cold spot – as life, too, was cold and hard; yet in the spring the daisies will come, and the thrushes will sing on the bough.

Built at first of seasoned wood, kept out of the weather under cover, repainted, and taken care of, the waggon lasts a lifetime. Many times repaired, the old ship outlasts its owner – his name on it is painted out. But that step is not taken for years: there seems to be a superstitious dislike to obliterating the old name, as if the dead would resent it, and there it often remains till it becomes illegible. Sometimes the second owner, too, goes, and the name fresh painted is that of the third. When at last it becomes too shaky for farm use, it is perhaps bought by some poor working haulier, who has a hole cut in the bottom with movable cover, and uses it to bring down flints from the hills to mend the roads. But if any of the old folk live, they will not sell the ancient vessel: it stands beside the rickyard under the elms till the rain rots the upper work, and it is then broken up, and the axletree becomes the top bar of a stile.

Each field has its characteristic stile – or rather two, one each side (at the entrance and exit of the footpath), and these are never alike. Walking across the fields for a couple of miles or more, of all the stiles that must of necessity be surmounted no two are similar. Here is one well put together – not too high, the rail not too large, and apparently an ideal piece of workmanship; but on approaching, the ground on the opposite side drops suddenly three or four feet – at the bottom is a marshy spot crossed by a narrow bridge of a single stone, on which you have to be careful to alight, or else plunge ankle-deep in water. If clever enough to drop on the stone, it immediately tilts up slightly, for, like the rocking stones of Wales, it is balanced somewhere, and has a see-saw motion well calculated to land the timid in the ditch.

The next is approached by a line of stepping stones – to avoid the mud and water – whose surfaces are so irregular as barely to afford a footing. The stile itself is nothing – very low and easy to pass: but just beyond it a stiff, stout pole has been placed across to prevent horses straying, and below that a couple of hurdles are pitched to confine the sheep. This is almost too much; however, by patience and exertion, it is managed.

Then comes a double mound with two stiles – one for each ditch – made very high and intended for steps; but the steps are worn away, and it is something like climbing a perpendicular ladder. Another has a toprail of a whole tree, so broad and thick no one can possibly straddle it, so some friend of humanity has broken the second rail, and you creep under. Finally comes a steep bank, six or seven feet high, with rude steps formed of the roots of trees worn bare by iron-tipped boots, and of mere holes in which to put the toe. At the top the stile leans forward over the precipice, so that you have to suspend yourself in mid-air. Fortunately, almost every other one has a gap worn at the side just large enough to squeeze through after coaxing the briars to yield a trifle. For it is intensely characteristic of human nature to make gaps and shortcuts.

All the lads of the hamlet have a trysting-place at the crossroads, or rather crosslanes, where there is often an open waste space and a small clump of trees. If there is any mischief in the wind, there the council of war is sure to be held. There is a great rickyard not far distant, where in one of the open sheds is the thatcher's workshop.

He is a very pronounced character in his way, with his leathern pads for the knees that he may be able to bear lengthened contact against the wooden rungs of the ladder, his little club to drive in the stakes, his shears to snip off the edges of the straw round the eaves, his iron needle of gigantic size with which to pass the tar cord through when thatching a shed, and his small sharp billhook to split out his thatching stakes. These are of willow, cut from the pollard trees by the brook, and he sits on a stool in the shed and splits them into three or four with the greatest dexterity, giving his billhook a twist this way and then that, and so guiding the split in the direction required. Then holding it across his knee, he cuts the point with a couple of blows and casts the finished stake aside upon the heap.

A man of no little consequence is the thatcher, the most important perhaps of the hamlet craftsmen. He ornaments the wheat ricks with curious twisted tufts of straw. But he does not put the thatch on the wheat half so substantially as formerly, because now only a few remain the winter – the thatch is often hardly on before it is off again for the thrashing machine – for the 'sheening', as they call it. On the hayricks, which stand longer, he puts better work, especially on the southern and western sides or angles, binding it down with a crosswork of bonds to prevent the gales which blow from those quarters unroofing the rick.

It is said to be an ill wind that blows nobody any good: now the wind never blew that was strong enough to please the thatcher. If the hurricane roughs up the straw on all the ricks in the parish, unroofs half a dozen sheds, and does not spare the gables of the houses, why he has work for the next two months. He is attended by a man to carry up the 'yelms', and two or three women are busy 'yelming' – i.e. separating the straw, selecting the longest and laying it level and parallel, damping it with water, and preparing it for the yokes. These yokes must be cut from boughs that have grown naturally in the shape wanted, else they are not tough enough. A tough old chap, too, is the thatcher, a man of infinite gossip, well acquainted with the genealogy of every farmer, and, indeed, of everybody from Dan to Beersheba, of the parish.

The memory of the smugglers is not yet quite extinct. The old men will point out the route they used to follow, and some of the places where they are said to have stored their contraband goods. Smuggling suggests the sea, but the goods landed on the beach had afterwards to be conveyed inland for sale, so that the hamlet, though far distant from the shore, has its traditions of illicit trade. The route followed was a wild and unfrequented one, and the smugglers appear to have kept to the downs as much as possible. More than one family – well-to-do for the hamlet or village, where a small capital goes a long way – are said to have originally derived their prosperity from assisting the storage or disposal of smuggled goods; and the sympathies of the hamlet would be with the smugglers still.

The old folk, too, talk of having the ague, and say that it was quite common in their early days; but it is rare to hear of a case now. Possibly the better drainage of the fields and the better food and lodging enjoyed by the labourers have something to do with this. There are, of course, no scientific or precise data for exact comparison; but, judging from the traditions transmitted down, the hamlet is much more healthy at the present day than it was in the olden times.

SEVEN

The Farmhouse

THE STREAM, after leaving the village and the washpool, rushes swiftly down the descending slope, and then entering the meadows, quickly loses its original impetuous character. Not much more than a mile from the village it flows placidly through meads and pastures, a broad, deep brook, thickly fringed with green flags bearing here and there large yellow flowers. By some old thatched cattle sheds and rickyards, overshadowed with elm trees, a strong bay or dam crosses it, forcing the water into a pond for the cattle, and answering the occasional purpose of a ford; for the labourers in their heavy boots walk over the bay, though the current rises to the instep. They call these sheds, some few hundred yards from the farmhouse, the 'Lower Pen'. Wick Farm – almost every village has its outlying 'wick' – stands alone in the fields. It is an ancient rambling building, the present form of which is the result of successive additions at different dates, and in various styles.

When a homestead like this has been owned and occupied by the same family for six or seven generations, it seems to possess a distinct personality of its own. A history grows up round about it; memories of the past accumulate, and are handed down fresh and green, linking today and seventy years ago as if hardly any lapse of time had intervened. The inmates talk familiarly of the 'comet year' as if it were but just over; of the days when a load of wheat was worth a little fortune; of the great snows and floods of the previous century. They date events from the year when the Foremeads were purchased and added to the patrimony, as if that transaction, which took place ninety years before, was of such importance that it must necessarily be still known to all the world.

The house has somehow shaped and fitted itself to the character of the dwellers within it: hidden and retired among trees, fresh and green with cherry and pear against the wall, yet the brown thatch and the old bricks subdued in tone by the weather. This individuality extends to the furniture; it is a little stiff and angular, but solid, and there are nooks and corners – as the window-seat – suggestive of placid repose: a strange opposite mixture throughout of flowery peace and silence, with an almost total lack of modern conveniences and appliances of comfort – as though the sinewy vigour of the residents disdained artificial ease.

In the oaken cupboards – not black, but a deep tawny colour with age and frequent polishing – may be found a few pieces of old china, and on the table at teatime, perhaps, other pieces, which a connoisseur would tremble to see in use, lest a clumsy arm should shatter their fragile antiquity. Though apparently so little valued, you shall not be able to buy these things for money – not so much because their artistic beauty, but because of the instinctive clinging to everything old, characteristic of the place and people. These have been there of old time: they shall remain still. Somewhere in the cupboards, too, is a curiously carved piece of iron, to fit into the hand, with a front of steel before the fingers, like a skeleton rapier guard; it is the ancient steel with which, and a flint, the tinder and the sulphur match were ignited.

Up in the lumber room are carved oaken bedsteads of unknown age; linen presses of black oak with carved panels, and a drawer at the side for the lavender bags; a rusty rapier, the point broken off; a flintlock pistol, the barrel of portentous length, and the butt weighted with a mace-like knob of metal, wherewith to knock the enemy on the head. An old yeomanry sabre lies about somewhere, which the good man of the time wore when he rode in the troop against the rioters in the days of machine burning – which was like civil war in the country, and is yet recollected and talked of. The present farmer, who is getting just a trifle heavy in the saddle himself, can tell you the names of labourers living in the village whose forefathers rose in that insurrection. It is a memory of the house how one of the family paid 40*l* for a substitute to serve in the wars against the French.

The mistress of the household still bakes a batch of bread at home in the oven once now and then, priding herself that it is never 'dunch', or heavy.

She makes all kinds of preserves and wines too – cowslip, elderberry, ginger – and used to on paper and baked by exposure to the sun's rays only. She has a bitter memory of some money having been lost to the family sixty years ago through roguery, harping upon it as a most direful misfortune; the old folk, even those having a stocking or a teapot well filled with guineas, thought a great deal of small sums. After listening to a tirade of this kind, in the belief that the family were at least half-ruined, it turns out to be all about 100*l*. Her grandmother after marriage travelled home on horseback behind her husband; there had been a sudden flood, and the newly married couple had to wait for several hours till the waters went down before they could pass. Times are altered now.

Since this family dwelt here, and well within what may be called the household memory, the very races of animals have changed or been supplanted. The cows in the fields used to be longhorns, much more hardy, and remaining in the meadows all the winter, with no better shelter than the hedges and bushes afforded. Now the shorthorns have come, and the cattle are housed carefully. The sheep were horned – up in the lumber room two or three horns are still to be found. The pigs were of a different kind, and the dogs and poultry. If the race of men have not changed they have altered their costume; the smock-frock lingered longest, but even that is going.

Some of the old superstitions hung on till quite recently. The value of horses made the arrival of foals an important occasion, and then it was the custom to call in the assistance of an aged man of wisdom – not exactly a wizard, but something approaching it nearly in reputation. Even within the last fifteen years the aid of an ancient like this used to be regularly invoked in this neighbourhood; in some mysterious way his simple presence and goodwill – gained by plentiful liquor – was supposed to be efficacious against accident and loss. The strangeness of the business was in the fact that his patrons were not altogether ignorant or even uneducated – they merely carried on the old custom, not from faith in it, but just because it was the custom. When the wizard at last died nothing more was thought about it. Another ancient used to come round once or twice in the year, with a couple of long ashen staves, and the ceremony performed by him consisted in dancing these two sticks together in a fantastic manner to some old rhyme or story.

The parlour is always full of flowers – the mantelpiece and grate in spring quite hidden by fresh green boughs of horse chestnut in bloom, or with lilac, bluebells, or wild hyacinths; in summer nodding grasses from the meadows, roses, sweet briar; in the autumn two or three great apples, the finest of the year, put as ornaments among the china, and the corners of the looking glass decorated with bunches of ripe wheat. A badger's skin lies across the back of the armchair; a fox's head, the sharp white tusks showing, snarls over the doorway; and in glass cases are a couple of stuffed kingfishers, a polecat, a white blackbird, and a diver – rare here – shot in the mere hard by.

On the walls are a couple of old hunting pictures, dusky with age, but the crudity of the colours by no means toned down, or their rude contrast moderated: bright scarlet coats, bright white horses, harsh green grass, prim dogs, stiff trees, human figures immovable in tight buckskins; running water hard as glass, the sky fixed, the ground all too small for the grouping, perspective painfully emphasised, so as to be itself made visible; the surface everywhere 'painty' – in brief, most of the possible faults compressed together, and proudly fathered by the artist's name in full.

One representing a meet, and the other full cry, the pack crossing a small river; the meet still and rigid, every horseman in his place – not a bit jingling, or a hoof pawing, or anything in motion. Now the beauty of the meet, as distinct from a drilled cavalry troop, is its animation: horses and riders moving here and there, gathering together and spreading out again, new-comers riding smartly up, in continuous freshness of grouping, and constant relief to the eye. The other – in full cry – all polished and smooth and varnished as when they left the stable; horses with glossy coats, riders upright and fatigueless, dogs clean, and not a sign of poaching on the turf. The dogs are coming out of the water with their tails up and straight – dogs as they trail their flanks out of a brook always, in fact, droop their tails, while their bodies look smaller and the curves project, because the water lays the hair flat to the body till several shakes send it out again. Not a speck on a top boot, not a coat torn by a thorn, and the horses as plump as if fresh from their mangers, instead of having worked it down. Not a fleck of foam; the sun, too, shining, and yet no shadow – all glaring. And, despite of all, deeply interesting to those who know the countryside and have a feeling knowledge of its hunting history.

For the horses are from life, and the men portraits; the very hedges and brooks faithful – in ground plan, at least. The costume is true to a thread, and all the names of the riders and some of the hounds are written underneath. So that a hunter sees not the crude colour or faulty drawing, but what it is intended to represent. Under its harshness there is the poetry of life. But looking at these pictures, the reflection will still arise how few really truthful hunting scenes we have on canvas in this the country of hunting. The best are so conventional, and have too much colour. All nature in the season is toned down and subdued – the gleaming red and bright yellows of the early autumn leaves soaked and soddened to a dull brown; the sky dark and louring – if it is bright there is frost; the glossy coat of the horses, and the scarlet, or what coloured cloth it may be, of the riders deadened by rain and the dewdrops shaken from the bushes. Think for a moment of a finish as it is in reality, and not in these gaudy, brilliant colour-studies.

A thick mist clings in the hollow there by the osier bed where the pack have overtaken the fox, so that you cannot see the dogs. Beyond, the contour of the hill is lost in the cloud trailing over it; the foreground towards us shows a sloping ploughed field, a damp brown, with a thin mist creeping along the cold furrows. Yonder, three vague and shadowy figures are pushing laboriously forward beside the leafless hedge; while the dirt-spattered bays hardly show against its black background and through the mist. Some way behind, a weary grey – the only spot of colour, and that dimmed – is gamely struggling – it is not leaping – through a gap beside a gaunt oak tree, whose dark buff leaves yet linger. But out of these surely an artist who dared to face Nature as she is might work a picture.

The year really commences at Wick farmhouse immediately before the autumn nominally begins – nominally, because there is generally a sense of autumn in the atmosphere before the end of September. Just about that time there comes a slackening of the work requiring earnest personal supervision. When the yellow corn has been cut and carted, and the thrashing machine has prepared a sample for the markets – when the ricks are thatched, and the steam plough is tearing up the stubble – then the farmer can spare a day or so free from the anxieties of harvest. There is plenty of work to be done; in fact, the yearly rotation of labour may be said to begin in the autumn too, but it does not demand such hourly attention. It is the season for picnics – while the sun is yet warm and the sward dry – on the downs among the great hazel copses, or the old entrenchment, with its view over a vast landscape, dimmed, though, by yellow haze, or by the shallow lake in the vale.

With the exception of knocking over a young rabbit now and then for household use, the farmer, even if he is independent of a landlord, as in this case, does not shoot till late in the year. Old-fashioned folk, though not in the least constrained to do so, still leave the first pick of the shooting to some neighbouring landowner between whose family and their own friendly relations have existed for generations. It is true that the practice becomes rarer yearly as the old style of men die out and the spirit of commerce is imported into rural life: the rising race preferring to make money of their shooting by letting it, instead of cultivating social ties.

At Wick, however, they keep up the ancient custom, and the neighbouring squire takes the pick of the wing-game. They lose nothing for their larder through this arrangement – receiving presents of partridges and pheasants far exceeding in number what could possibly be killed upon the farm itself; while later in the year the boundaries are relaxed on the other side, and the farmer kills his rabbit pretty much where he likes, in moderation.

He is seldom seen without a gun on his shoulder from November till towards the end of January. No matter whether he strolls to the arable field, or down the meadows, or across the footpath to a neighbour's house, the inevitable double-barrel accompanies him. To those who live much out of doors a gun is a natural and almost a necessary companion, whether there be much or little to shoot; and in this desultory way, without much method or set sport, he and his friends, often meeting and joining forces, find sufficient ground game and wildfowl to give them plenty of amusement. When the hedges are bare of leaves the rabbit burrows are ferreted: the holes can be more conveniently approached then, and the frost is supposed to give the rabbits a better flavour.

About Christmas time, half in joke and half in earnest, a small party often agree to shoot as many blackbirds as they can, if possible to make up the traditional twenty-four for a pie. The blackbird pie is, of course, really an occasion for a social gathering, at which cards and music are forthcoming. Though blackbirds abound in every hedge, it is by no means an easy task to get the required number just when wanted. After January the guns are laid aside, though some ferreting is still going on.

The better class of farmers keep hunters, and ride constantly to the hounds; so do some of the lesser men who 'make' hunters, and ride not only for pleasure but possible profit from the sale. Hunting is, to a considerable extent, a matter of locality. In some districts it is the one great winter amusement, and almost every farmer who has got a horse rides more or less. In others which are not near the centres of hunting, it is rather an exception for the farmers to go out. On and near the downs coursing hares is much followed. Then towards the spring, before the grass begins to grow long, comes the local steeplechase – perhaps the most popular gathering of the year. It is held near some small town, often rather a large village than a town, where it would seem impossible to get a hundred people together. But it happens to be one of the fixed points, so to say, in a wide hunting district, and is well known to every man who rides a horse within twenty miles.

Numerous parties come to the race ground from the great houses of the neighbourhood. The labouring people flock there *en masse*; some farmers lend waggons and teams to the labourers that they may go. An additional – a personal – interest attaches to many of the races because the horses are local horses, and the riders known to the spectators. Some of these meetings are movable; they are held near one town one year and another the next, so as to travel round the whole hunting district – returning, say, the fourth year to the first place. Most of the market towns of any importance have

their annual agricultural show now, which is well attended.

In the spring comes the rook shooting; the date varies a week or so according to the season, whether it has been mild and favourable or hard and late. This still remains a favourite occasion for a party. Sheep shearing in sheep districts, as the downs, is also remembered; some of the old folk make much of it; but as a general rule this ancient festival has fallen a good deal into disuse. It is not made the grand feast it once was for master and man alike – at least, not in these parts. With the change that has come across agriculture at large a variation has taken place in the life of the people. New festivals, and of a different character, have sprung up.

The most important of these is the annual auction on the farm: the system of selling by auction which has become so widely diffused has, indeed, quite revolutionised agriculture in many ways. Where the farm is celebrated for a special breed of sheep, the great event of the year is the annual auction at home of ram lambs. Where the farm is famous for cattle, the chief occasion is the yearly sale of young shorthorns. And recently, since steam plough and artificial manure and general high pressure have been introduced, many large arable farmers sell their corn crops standing. The purchaser pays a certain price for the wheat as it grows, reaps it when ripe, and makes what profit he can.

In either case the auctioneer is called in, a dinner is prepared, and everybody who likes to come is welcome. If there happens to be a great barn near the homestead it is usually used for the dinner. The marquee has yet to be invented which will keep out a thunderstorm – that common interruption of country meetings – like an old barn. But barns are not always available, and a tent is then essential. Though the spot may be lonely and several miles from a town or station, a large number of persons are sure to be there; and if it is an auction of sheep or cattle with a pedigree, many of them will be found to have come from the other end of the kingdom, and sometimes agents are present from America or the colonies. Much time is consumed in an examination of the stock, and then the dinner begins – at least two hours later than was announced. But this little peculiarity is so well understood by all interested as to cause no inconvenience.

Scarcely any ale is to be seen; it is there if asked for; but the great majority now drink sherry. The way in which this wine has supplanted the old-fashioned October ale is remarkable, and a noticeable sign of the times. At home the farmer may still have his foaming mug, but whenever farmers congregate together on occasions like this, sherry is the favourite. When calling at the inns in the towns on market days – much business is transacted at the inns – spirits are usually taken, so that ale is no longer the characteristic country liquor. With the sherry, cigars are handed round – another change. It is true the elderly men stick to their long clay pipes, and it is observable that some of the younger after a while go back to the yard of clay; but on the whole the cigar is now the proper thing.

Then follow a couple of toasts, the stockowner's and auctioneer's – usually short – and an adjournment takes place – if it be stock, to the yards; if corn, the cloth is cleared of all but the wine, and the sale proceeds there and then. In either case the sherry and the cigars go round – persons being employed to press them freely upon all; and altogether a very jovial afternoon is spent. Some of the company do not separate till long after the conclusion of the sale: the American or colonial agent perhaps stays a night at the farmstead. In the house itself there is all this time yet more liberal hospitality proffered: it is quite open-house hospitality, master and mistress vying in their efforts to make everyone feel at home. These gatherings do much to promote a friendly spirit in the neighbourhood.

In the summer the farmer is too much occupied to think of amusement. It is a curious fact that very few really downright country people care for fishing; a gun and a horse are as necessary as air and light, but the rod is not a favourite. There seems to be greater enthusiasm than ever about horses; whether people bet or not, they talk and think and read more of horses than they ever did before.

In this locality Clerk's Ale, which used to be rather an event, is quite extinct. The Court Leet is still held, but partakes slightly of the nature of a harmless farce. The lord of the manor's court is no terror now. A number of gentlemen, more for the custom's sake than anything, sit in solemn conclave to decide whether or no an old pollard tree may be cut down, how much an old woman shall pay in quit-rent for her hovel, or whether there was or was not a gateway in a certain hedge seventy years ago. However, it brings neighbours together, and causes the inevitable sherry to circulate briskly.

The long summer days begin very early at Wick. About half-past two of a morning in June a faint twittering under the eaves announces that the swallows are awaking, although they will not commence their flight for a while yet. At three o'clock the cuckoo's call comes up from the distant meadows, together with the sound of the mower sharpening his scythe, for he likes to work while the dew is heavy on the grass, both for coolness and because it cuts better. He gets half a day's work done before the sun grows hot, and about eight or nine o'clock lies down under the hedge for a refreshing nap. Between three and four the thrushes open song in the copse at the corner of the Homefield, and soon a loud chorus takes up their ditty as one after the other joins in.

Then the nailed shoes of the milkers clatter on the pitching of the courtyard as they come for their buckets; and immediately afterwards stentorian voices may be heard in the fields bellowing 'Coom up! ya-hoop!' to which the cows, recognising the well-known call, respond very much in the same tones. Slowly they obey and gather together under the elms in the corner of the meadow, which in summer is used as the milking place. About five or half-past another clattering tells of the milkers' return; and then the dairy is in full operation. The household breakfasts at half-past six or thereabouts, and while breakfast is going on the heavy tramp of feet may be heard passing along the roadway through the rickyard – the haymakers marching to the fields. For the next two hours or so the sounds from the dairy are the only interruption of the silence; then come the first waggons loaded with hay, jolting and creaking, the carter's lads shouting, 'Woaght!' to the horses as they steer through the gateway and sweep round, drawing up under the rick.

Between eleven and twelve the waggons cease to arrive – it is luncheon time: the exact time for luncheon varies a quarter of an hour or twenty minutes, or more, according to the state of the work. Messengers come home for cans of beer, and carry out also to the field wooden 'bottles' – small barrels holding a gallon or two. After a short interval work goes on again till nearly four o'clock, when it is dinnertime. One or two labourers, deputed by the rest and having leave and licence so to do, enter the farmhouse garden and pull up bundles of onions, lettuces, or radishes – sown over wide areas on purpose – and carry them out to the cart house, or wherever the men may be. If far from home, the women often boil a kettle for tea under the hedge, collecting dead sticks fallen from the trees. At six o'clock work is over: the women are allowed to leave half an hour or so previously, that they may prepare their husbands' suppers.

As the sunset approaches the long broad dusty road loses its white glare, and yonder by the hamlet a bright glistening banner reflects the level rays of the sun with dazzling sheen; it is the gilding on the swinging wayside sign transformed for the moment from a wooden board rudely ornamented with a gilt sun, all rays and rotund cheeks, into a veritable oriflamme.

There the men will assemble by and by, on the forms about the trestle table, and share each other's quarts in the fellowship of labour. Or perhaps the work may be pressing, and the waggons are loaded till the white owl noiselessly flits along the hedgerow, and the round moon rises over the hills. Then those who have stayed to assist find their supper waiting for them in the brewhouse, and do it ample justice.

Once during the morning, while busy in the hayfield, not so much with his hands as his eyes, watching that the 'wallows' may be turned over properly, and the 'wakes' made at a just distance from each other, that the waggon may pass easily between, the farmer is sure to be summoned home with the news of a swarm of bees. If the work be pressing, they must be attended to by deputy; if not, he hurries home himself; for although in these days bee-keeping is no longer what it used to be, yet the old-fashioned folk take a deep interest in the bees still. They tell you that 'a swarm in May is worth a load of hay; a swarm in June is worth a silver spoon; but a swarm in July is not worth a fly' – for it is then too late for the young colony to store up a treasure of golden honey before the flowers begin to fade at the approach of autumn.

It is noticeable that those who labour on their own land (as at Wick) keep up the ancient customs much more vigorously than the tenant who knows that he is liable to receive a notice to quit. And farms, for one reason or another, change tenants much more frequently now than they used to do. Here at Wick the owner feels that every apple tree he grafts, every flower he plants, returns not only a money value, but a joy not to be measured by money. So the bees are carefully watched and tended, as the blue tomtits find to their cost if they become too venturesome.

These bold little bandits will sometimes make a dash for the hive, alighting on the miniature platform before the entrance, and playing havoc with the busy inmates. If alarmed they take refuge in the apple trees, as if conscious that the owner will not shoot them there, since every pellet may destroy potential fruit by cutting and breaking those tender twigs on which it would presently grow. It is a pleasant sight in autumn to see the room devoted to the honey – great broad milk tins full to the brim of the translucent liquid, distilling slowly from pure white comb, from the top of whose cells the waxen covering has been removed.

All the summer through fresh beauties, indeed, wait upon the owner's footsteps. In the spring the mowing-grass rises thick, strong, and richly green, or hidden by the cloth-of-gold thrown over it by the buttercups. He knows when it is ready for the scythe without reference to the almanac, because of the brown tint which spreads over it from the ripening seeds, sometimes tinged with a dull red, when the stems of the sorrel are plentiful. At first the aftermath has a trace of yellow, as if it were fading; but a shower falls, and fresh green blades shoot up. Or, passing from the hollow meads up on the rising slopes where the plough rules the earth, what so beautiful to watch as the wheat through its various phases of colour?

First green and succulent; then, presently, see a modest ear comes forth with promise for the future. By and by, when every stalk is tipped like a sceptre, the lower stalk leaves are still green, but the stems have a faint bluish tinge, and the ears are paling into yellow. Next the white pollen – the bloom – shows under the warm sunshine, and then the birds begin to grow busy among it. They perch on the stalk itself – it is at that time strong and stiff enough to uphold their weight, one on a stem – but not now for mischief. You may see the sparrow carry away with him caterpillars for his young upon the house-top hard by; later on, it is true, he will revel on the ripe grains.

Yesterday you came to the wheat, and found it pale like this (it seems but twenty-four hours ago – it is really only a little longer); today, when you look again, lo! there is a fleeting yellow already on the ears. They have so quickly caught the hue of the bright sunshine pouring on them. Yet another day or two, and the faint fleeting yellow has become fixed and certain, as the colours are deepened by the great artist. Only when the wind blows and the ears bend in those places where the breeze takes most, it looks paler because the under part of the ear is shown and part of the stalk. Finally comes that rich hue for which no exact similitude exists. In it there is somewhat of the red of the orange, somewhat of the tint of bronze, and somewhat of the hue of maize; but these are poor words wherewith to render fixed a colour that plays over the surface of this yellow sea, for if you take one, two, or a dozen ears you shall not find it, but must look abroad, and let your gaze travel to and fro. Nor is every field alike; here are acres and acres more yellow, yonder a space whiter, beyond that a slope richly ruddy, according to the kind of seed that was sown.

Out of the depths of what to it must seem an impenetrable jungle, from visiting a flower hidden below, a humblebee climbs rapidly up a stalk a yard or two away while you look, and mounting to the top of the ear, as a post of vantage clear of obstructions, sails away upon the wind.

'We be all jolly vellers what vollers th' plough!' But not to listen to, and take literally according to the letter of the discourse. It runs something like this the seasons through as the weather changes: 'Terrible dry weather this here to be sure; we got so much work to do uz can't get drough it. The fly be swarming in the turmots – the smut be on the wheat – the wuts be amazing weak in the straw. Got a fine crop of wheat this year, and prices be low, so uz had better drow it to th' pigs. Last year uz had no wheat fit to speak on, and prices was high. Drot this here wet weather! the osses be all in the stable eating their heads off, and the chaps be all idling about and can't do no work: a pretty penny for wages and not a job done. Them summer ricks be all rotten at bottom. The ploughing engine be stuck fast up to the axle, the land be so soft and squishey. Us never gets no good old frosts now, like they used to have. Drot these here frosty mornings! a-cutting up everything. There'll be another rate out soon, a' reckon. Us had better give up this here trade, neighbour!'

And so on for a thousand and one grumbles, fitting into every possible condition of things, which must not, however, be taken too seriously; for of all other men the farmer is the most deeply attached to the labour by which he lives, and loves the earth on which he walks like a true autochthon. He will not leave it unless he is suffering severely.

EIGHT

Birds of the Farmhouse

WICK FARMHOUSE IS THATCHED, and has many gables hidden with ivy. In these broad expanses of thatch, on the great 'chimney-tuns', as country folk call them, and in the ivy, tribes of birds have taken up their residence. The thatch has grown so thick in the course of years by the addition of fresh coats that it projects far from the walls and forms wide, far-reaching eaves. Over the cellar the roof descends within three or four feet of the ground, the wall being low, and the eaves here cast a shadow with the sun nearly at the zenith.

On the higher parts of the roof, especially round the chimneys, the starlings have made their holes, and in the early summer are continuously flying to and fro for their young, who never cease crying for food the whole day through. A tall ash tree stands in the hedgerow, about fifty yards from the house. On this tree, which is detached, so that they can see all round, the starlings perch before they come to the roof, as if to reconnoitre, and to exchange pourparlers with their friends already on the roof; for if ever birds talk together starlings do. Many birds utter the same notes over and over again; others sit on a branch and sing the same song, as the thrush; but the starling has a whole syllabary of his own, every note of which evidently has its meaning, and can be varied and accented at pleasure.

His whistle ranges from a shrill, piercing treble to a low, hollow bass; he runs a complete gamut, with 'shakes', trills, tremulous vibrations, every possible variation. He intersperses a peculiar clucking sound, which seems to come from the depths of his breast, fluttering his wings all the while against his sides as he stands bolt upright on the edge of the chimney. Other birds seem to sing for the pure pleasure of singing, shedding their notes broadcast, or at most they are meant for a mate hidden in the bush. The starling addresses himself direct to his fellows: I think I may say he never sings when alone, without a companion in sight. He literally speaks to his fellows. I am persuaded you may almost follow the dialogue and guess the tenor of the discourse.

A starling is on the chimney top; yonder on the ash tree are four or five of his acquaintance. Suddenly he begins to pour forth a flood of eloquence – facing them as he speaks: Will they come with him down to the field where the cows are grazing? There will be sure to be plenty of insects settling on the grass round the cows, and every now and then they tear up the herbage by the roots and expose creeping things. 'Come,' you may hear him say, modulating his tones to persuasion, 'come quickly; you see it is a fresh piece of grass into which the cows have been turned only a few hours since; it was too long for us before, but where they have eaten we can get at the ground comfortably. The water wagtail is there already; he always accompanies the herd, and will have the pick and choice of everything. Or what do you say to the meadow by the brook? The mowers have begun, and the swathe has fallen before their scythes; there are acres of ground there which we could not touch for weeks; now it is open, and the place is teeming with good food. The finches are there as busy as may be between the swathes – chaffinch and greenfinch, hedge sparrow, thrushes, and blackbirds too. Are you afraid? Why, no one shoots in the middle of a summer's day. Still irresolute? (with an angry shrillness). Will you or will you not? (a sharp, short whistle of interrogation). You are simply idiots (finishing with a scream of abuse). I'm off!'

Seeing him start, the rest follow at once, jealous lest he should enjoy these pleasures alone. As he flies every few minutes he closes his wings, so that for half a dozen yards he shoots like an arrow through the air; then rapidly uses them, and again closes and shoots forward, all the time keeping a level straight course, going direct to his object.

The starlings that breed in the roof, though they leave the place later on and congregate in flocks roosting in trees, still come back now and then to revisit their homes, especially as the new year opens, when they alight on the house frequently and consult on the approaching important period of nesting. If you should be sitting near a window close under the roof where they are busy, reading a book, with the summer sunshine streaming in, now and then a flash like lightning will pass across the page. It is a starling rapidly vibrating his wings before he perches on the thatch; the swift succession of light and shadow as the wings intercept the rays of the sun causes an impression on the eye like that left by a flash of lightning. They are beautiful birds: on their plumage, when seen quite close, the light plays in iridescent gleams.

Upon the roof of the old farmstead, too, the chirp of the sparrow never ceases the livelong day. It is amusing to see these birds in the nesting season carrying up long straws or feathers – towing their burden through the air with evident labour. These they sometimes drop just as they arrive at their destination. Eager to utter a chirp to their mates, they open their beaks, and away floats the feather, but they catch it again before it reaches the ground. Fluffy feathers are great favourites. The fowls, as they fly up to roost on the beams in the sheds, beat out feathers from their clumsy wings; these lie scattered on the ground, marking the spot. These roosting-places are magazines from which the small birds draw their supplies for domestic purposes. The sparrows have their nests in lesser holes in the thatch; sometimes they use a swallow's nest built of mortar under the eaves, to which the owners have not returned.

The older folk still retain some faint superstitions about swallows, looking upon them as semi-consecrated and not to be killed or interfered with. They will not have their nests knocked down. If they do not return to the eaves but desert their nests, it is a sign of misfortune impending over the household. So, too, if the rooks quit the rookery, or the colonies of bees in the hives on the sunny side of the orchard decay and do not swarm, but seem to die off, it is an evil omen. If at night a bird flutters against the windowpane in the darkness – as they will sometimes in a great storm of wind, driven, perhaps, from their roosting-places by the breaking of the boughs, and attracted by a light within – the knocking of their wings betokens that something sad is about to happen. If an invalid asks for a pigeon – taking a fancy to a dish of pigeons to eat – it is a sign either of coming dissolution or of extreme illness.

But the swallows rarely fail to come in the spring, and soon begin to repair their nests or build new ones with mortar from the roads; a rainy day is very useful to them, and they alight at the edge of the puddles, finding the mud already mixed and tempered for them there. In such weather they will fly backwards and forwards by the side of a hedge for a length of time, skimming just above the grass, when, looking down on them instead of up at them, the white bar across the lower part of the body just before the tail forks is very noticeable. The darker feathers have a glossy bluish tinge on the black. They seem fond of flying round near horses and cattle, as if insects were more numerous near animals. While driving on a sultry day I have watched a swallow follow the horse for a mile or more.

It is a pleasant sight to watch them gliding just above the surface of smooth water, dipping every now and then. Once, while observing some swallows flying over a lake, on a windy day, when there were waves of some size, I saw a swallow struck by the crest of a wave and overwhelmed. It was about twenty yards from a lee shore, and the bird floated on the water, rising and sinking with the waves till they threw it on the bank. It was much exhausted, but when placed on a stone in the warm sunshine soon recovered and flew off.

As another proof that, quick as they are on the wing, they do not always judge their position or course precisely, I know a case where a swallow, in less than ten yards after leaving her nest under the eaves of a house, flew with great force against a door in the garden wall painted a dull blue. The beak was partly broken and the bird completely stunned: she died in a few minutes. There was some one in the garden close by at the time: his presence may have frightened the swallow; yet they are not usually timid where their nests are undisturbed. Perhaps in her hurry the dull blue colour of the gate may have deceived her sight; but she must have travelled that way a hundred times before.

Swallows frequently come down the great chimneys at the farmhouse and are found in the rooms, but are always allowed to escape from the window. Swallows are said not to perch; but I have seen them repeatedly perch on those sticks which, where the thatch has somewhat decayed, project a few inches above the roof tree. Sometimes a row of half a dozen may be observed settled on the roof here. You may see them, too, perch on the topmost boughs of the tall damson trees in the orchard; and again, later in the autumn, after nesting is over, they assemble in hundreds – one might almost say thousands – in the withy bed by the brook, settling on the slender willow wands. There they twitter together for an hour or more every evening. They can rise without the slightest difficulty from the ground, if it is level and not encumbered with grass, as from the surface of the roads. On dull, cold days they settle on the house more frequently than when it is bright and sunny.

At one end of the farmhouse, which is an irregular building, there is a quiet gable, and in it a casement arched over by the thatch, and shaded by a thick growth of ivy. The casement is low, and not more than eight or nine feet from the ground; the ivy has climbed the wall, it has spread, too, over the massive wall of the garden which just there abuts upon the house, so that there is a secluded corner formed by the angle. Here some time ago a number of logs of timber – oak, such as are sawn up into posts for field gateways – were left leaning half against the garden wall, half against the house, just under the window. There they have remained (there is never any hurry about things in the country) so long that the moss has begun to encase the lower portions. What with projecting thatch, the thick ivy, the timber thrown carelessly beneath, the lichen-grown garden wall, and a large bush of lilac in the angle, the place could hardly be more quiet, and is consequently a favourite resort of the birds.

Within reach from the window the swallows have their nests, and the sparrows their holes, on the right hand; within reach on the left hand, among the ivy, the water wagtail has built her nest year after year. The wagtail may always be seen about the place – now in the cowyards among the cattle, now in the rickyard, and even close to the door of the house, especially frequenting the courtyard in front of the dairy. As he flies he rises up and then sinks again, in a succession of undulations, now spreading the tail out and now closing it. On the ground he generally alights near water; he is continually jerking the tail up and down.

One spring a cuckoo came to this nest in the ivy close to the casement; she was seen flying near the house several times, and, being observed to visit the ivy-covered gable, was finally traced to the wagtail's nest. For several days in succession, and several times a day, the cuckoo came, and would doubtless have left an egg had not she been shot by a person who wanted a cuckoo to stuff.

It is difficult to understand upon what principle the cuckoo selected a nest thus placed. The ordinary considerations put forward as guiding birds and animals in their actions quite fail. Instinct would scarcely choose a spot so close to a house – actually on it; the desire of safety would not lead to it either, nor the idea of concealment. She might, no doubt, have found nests enough at a distance from houses, and much more likely to escape observation. Was there any kind of feeling that this particular wagtail was more likely to take care of the offspring than others?

I doubt the cuckoo's alleged total indifference to her young. They certainly linger in the neighbourhood of the nests which they have selected to deposit their eggs in. On another occasion a cuckoo used a wagtail's nest in a different part of the garden here – in some ivy that had grown round the decaying stump of an old fir tree. This bird was watched, but not interfered with; she came repeatedly, and was seen on the nest, and the egg observed. Afterwards a cuckoo sang continuously day after day on an ash tree close to the garden.

Lower down in the ivy, behind the logs of timber under the casement, the hedge sparrow builds every year; and on the wood itself where the trunks formed a little recess was a robin's nest. The hedge sparrow, unlike his noisy namesake, is one of the quietest of birds: he slips about in the hedges and bushes all round the garden so quietly and unobtrusively that unless you watch carefully you will not see him. Yet he does not seem shy, and if you sit still will come along the hawthorn within a yard.

In the thatch – under the eaves of the cellar, which are not more than four feet from the ground and come up to the ivy of the gable – the wren has a nest. Some birds seem always to make their nests in one particular kind of way, and generally in the same kind of tree or bush; robins, house sparrows, and starlings, on the other hand, adjust their nests to all sorts of places.

The window of a room in which I used to sleep overlooked the orchard, and there was a pear tree trained against the wall, some of the boughs of which came up to the windowsill. This pear tree acted as a ladder, up which the birds came. Pear trees are a good deal frequented by many birds; their rough bark seems to shelter numerous insects. The window was left open all night in the sultry summer weather, and presently a robin began to come in very early in the morning. Encouraged by finding that no one disturbed him, at last he grew bold enough to perch morning after morning on the rail at the foot of my bed. First he seemed to examine the inside of the window, then went on the floor, and, after a good look round, finally finished by sitting on the wooden framework for a few minutes before departing.

This went on some time; then a wren came too; she likewise looked to see if anything edible could be found in the window first. Old-fashioned windows often have a broad sill inside – the window frame being placed nearly at the outer edge of the wall, so that the thickness of the wall forms a recess, which is lined with board along the bottom. Now this wooden lining was decayed and drilled with innumerable holes by boring insects, which threw up tiny heaps of sawdust, as one might say, just as moles throw up mounds of earth where they tunnel. Perhaps these formed an attraction to the wren. She also frequently visited an old-fashioned bookcase, on the top which – it was very low – I often left some old worm-eaten folios and quartos, and may have occasionally picked up something there. Once only she ventured to the foot of the bed. After leaving the room she always perched on a thin iron projection which held the window open, and uttered her singularly loud notes, their metallic clearness seeming to make the chamber ring. Starlings often perched on the same iron slide, and sparrows continually; but only the robin and wren came inside. Tomtits occasionally entered and explored the same board-lining of the window, but no farther. They will, however, sometimes explore a room.

I know a parlour the window of which was partly overhung by a similar pear tree, besides which there were some shrubs just outside, and into this room, being quiet and little used, the tomtits ventured every now and then. I fancy the placing of flowers in vases on the table or on the mantelpiece attracts birds to rooms, if they are still. Insects visit the flowers; birds look for insects: and this room generally abounded with cut flowers. Entering it suddenly one day, a tomtit flew from side to side in great agitation, and then dropped on the floor and allowed me to pick it up without an effort to escape. The bird had swooned from fright, and was quite helpless – the eyes closed. On being placed outside the window, in five minutes it came to itself and flew off feebly. In this way birds may frequently become a prey to cats and hawks when to all appearance they might easily escape – becoming so overwhelmed with alarm as to lose the power of motion.

The robin is a most pugnacious creature. He will fight furiously with a rival; in fact, he never misses an opportunity of fighting. But he always chooses the very early morning for these encounters, and so escapes suspicion, except, of course, from people who rise early too. It is even said that the young cock robins, when they are full-grown, turn round on their own parents and fight with them vigorously. Neither is he a favourite with the upper class of cottagers – for there is an 'upper ten' even among cottagers – who have large fruit gardens. In these they grow quantities of currants for preserving purposes. The robin is accused of being a terrible thief of currants, and meets with scant mercy.

Sometimes while walking slowly along the footpath in a lane with hedges each side, a robin will dart out of the hawthorn and pick up a worm or grub almost under your feet; then in his alarm at your presence drop it, and rush back in a flutter. Other birds will do the same thing, from which it would seem that the old saying that the eye sees what it comes to see is as applicable to them as to human beings. Their eyes, ever on the watch for food, instantly detect a tiny creeping thing several yards distant, though concealed by grass; but the comparatively immense bulk of a man appears to escape notice till they fly almost up against it.

I fancy that the hive bee and some kindred insects have a special faculty of seeing colour at a distance, and that colours attract them. It can hardly be scent, because when flowers are placed in a room and the window left open, the wind generally blows strongly into the apartment, and odours will not travel against a breeze. It seems natural that in both cases the continual watch for certain things should enable bird and insect to observe the faintest indication. Slugs, caterpillars, and such creatures, too, in moving among the grass, cause a slight agitation of the grass blades; they lift up a leaf by crawling under it, or depress it with their weight by getting on it. This may enable the bird to detect their presence, even when quite hidden by the herbage, experience having taught it that when grass is moved by the wind broad patches sway simultaneously, but when an insect or caterpillar is the agent only a single leaf or blade is stirred.

At the farmhouse here, robins, wrens, and tomtits are always hanging about the courtyard, especially close to the dairy, where one or other may be constantly seen perched on the palings; neither do they scruple to enter the dairy, the brewhouse, or woodhouse adjacent, when they see a chance. The logs (for fuel) stored in the latter doubtless afford them insects from under the dead bark.

Among the most constant residents in the garden at Wick Farm are the song thrushes. They are the tamest of the larger birds; they come every morning right under the old bay window of the sitting room on the shady side of the house, where the musk plant has spread and covered the stone pitching for many yards, except just a narrow path paved with broad flagstones. The musk finds root in every interstice of the pitching, but cannot push up through the solid flat flags; a fungus, however, has attempted even that, and has succeeded in forcing a great stone, weighing perhaps fifteen or twenty pounds, from its bed, so that instead of being level it forms an inclined plane. The carpet of musk yields a pleasant odour; in one corner, too, the monkey plant grows luxuriously, and the grass of the green or lawn is for ever trying to encroach upon the paving. In the centre of the green is a bed of gooseberries and a cherry tree; and though the fruit is so close to the window, both thrush and blackbird make as free with it as if it was in the hedgerow.

The thrush, when he wishes to approach the house, flies first to the cover of these gooseberries; then, after reconnoitring a few minutes, comes out on the green and gradually works his way across it to the stone pitching, and so along under the very window. The blackbird comes almost as often to the lawn, but it is in a different way. His manner is that of a bold marauder, conscious that he has no right, and aware that a shot from an ambuscade may lay him low, but defiantly risking the danger. He perches first on a bush, or on the garden wall, under the sheltering boughs of the lime trees, at a distance of some twenty yards; then, waiting till all is clear, he makes a desperate rush for the fruit trees or the lawn. The moment he has succeeded in violently seizing some delicious morsel off he goes, uttering a loud chuckle – half as a challenge, half as a vent for his pent-up anxiety.

This peculiar chuckle is so well known by all the other birds as a note of alarm that every one in the garden immediately moves his position, if only a yard or two. When you are stealing down the side of the hedgerow, endeavouring to get near enough to observe the woodpecker in a tree, or with a gun to shoot a pigeon, the great anxiety is lest you startle a blackbird. If he thinks you have not seen him, he is cunning enough to slip out the other side noiselessly and fly down beside the hedge just above the ground for some distance. He then crosses the field to a hedge on the other side, and, just as he safely lands himself in a thick hawthorn bush a hundred yards away, defiantly utters his cry. The pigeon or the woodpecker will instantly glance round; but, the cry being at a distance, if you keep still a minute or two they will resume their occupation. But if you should disturb the blackbird on the side of the bank next you, where he knows you must have seen or heard him, or if he is obliged to come out on your side of the hedge, then he makes the meadow ring with his alarm-note, and immediately away goes pigeon or woodpecker, thrushes fly farther down the hedge, and the rabbits feeding in the grass lift up their heads and, seeing you, rush to their burrows. In this way the blackbird acts as a general sentinel.

He has two variations of this cry. One he uses when just about to change his feeding ground and visit another favourite corner across the field; it is as much as to say, 'Take notice, all you menials; I, the king of the hedge, am coming.' The other is a warning, and will very often set two or three other blackbirds calling in the same way whose existence till then was unsuspected. These calls are quite distinct from his song.

Sometimes, when sitting on a rail in the shade of a great bush – a rail placed to close a gap – I have had a blackbird come across the meadow and perch just above my head. Till the moment of alighting he was ignorant of my presence, and for a second the extremity of his astonishment literally held him speechless at his own temerity. The next – what an outcry and furious bustle of excitement to escape! So in the garden here he makes a desperate rush, seizes his prey, and off again twenty or thirty yards, exhibiting an amusing mixture of courage and timidity. This process he will repeat fifty times a day. No matter how terribly frightened, his assurance quickly returns, and another foray follows; so that you begin by thinking him the most cowardly and end by finding him the most impudent of birds.

I love the blackbird, and never weary of observing him. There is a bold English independence about him – an insolent consciousness of his own beauty. He must somehow have read Shakespeare, for he seems quite aware of his 'orange tawny bill' and deep black hue. He might really know that he figures in a famous ballad, and that four-and-twenty of his species were considered a dish to set before a king.

It is a sight to see him take his bath. In a meadow not far from the house here is a shallow but clear streamlet, running down a deep broad ditch overshadowed by tall hemlock and clogweed, arched over with willow, whose leaves when the wind blows and their underside is exposed give the hedge a grey tint, with maple and briar. Hide yourself here on a summer morning among the dry grass and bushes, and soon the blackbird comes to stand a minute on a stone which checks the tiny stream like a miniature rock, and then to splash the clear water over head and back with immense energy. He repeats this several times, and immediately afterwards flies to an adjacent rail, where, unfettered by boughs, he can preen his feathers, going through his toilet with the air of a prince. Finally, he perks his tail up, and challenges the world with the call already mentioned, which seems now to mean, 'Come and see Me; am I not handsome?'

On a warm June day, when the hedges are covered with roses and the air is sweet with the odour of mown grass, it is pleasant to listen to the blackbirds in the oaks pouring forth their rich liquid notes. There is no note so sweet and deep and melodious as that of the blackbird to be heard in our fields; it is even richer than the nightingale's, though not so varied. Just before noonday – between eleven and twelve – when the heat increases, he leaves the low thick bushes and moist ditches and mounts up into an oak tree, where, on a branch, he sits and sings. Then another at a distance takes up the burden, till by and by, as you listen, partly hidden in a gateway, four or five are thus engaged in the trees of a single meadow.

He sings in a quiet, leisurely way, as a great artist should – there is no haste, no notes thickening on notes in swift crescendo. His voice (so to speak) drops from him, without an effort, and is so clear that it may be heard at a long distance. It is not a set song; perhaps, in strict language, it is hardly a song at all, but rather a succession of detached notes with intervals between. Except when singing, the blackbird does not often frequent trees; he is a hedge-bird, though sometimes when you are looking at a field of green corn or beans one will rise out of it and fly to a tree – a solitary tree such as is sometimes seen in the midst of an arable field. At Wick Farm, sitting in the cool parlour, or in the garden under the shade of the trees you may hear him almost every morning in the meadows that come right up to the orchard hedge. That hedge is his favourite approach to the garden: he flies to it first, and gradually works his way along under cover till nearer the cultivated beds. Both blackbird and thrush are particularly fond of visiting a patch of cabbages in a shady, quiet corner: there are generally two or three there after the worms and caterpillars, and so forth.

The thrushes build in the garden in several places, especially in an ivy-hidden arbour – a wooden frame completely covered with ivy and creeping flowers. Close by is a thick box hedge, six feet high and nearly as much through, and behind this is a low-thatched tool house, where spades, mole traps, scythes, reaping hooks, and other implements are kept. Here lies a sarsen stone, hard as iron, about a foot thick, the top of which chances to be smooth and level. This is the thrush's favourite anvil.

He searches about under the ivy, under which the snails hide in their shells in the heat of the day, and brings them forth into the light. The shell is too large for his beak to hold it pincer-fashion, but at the entrance – the snail's doorway – he can thrust his bill in, and woe then to the miserable occupant! With a hop and flutter the thrush mounts the stone anvil, and there destroys his victim in workmanlike style. Up goes his head, lifting the snail high in the air, and then, smash! the shell comes down on the stone with all the force he can use. About two such blows break the shell, and he then coolly chips the fragments off as you might from an egg, and makes very few mouthfuls of the contents. On the stone and round about it lie the fragments of many such shells – relics of former feasts. Sometimes he will do this close to the bay window – if all is quiet – using the stone flags for an anvil, if he chances to find a snail hard by; but he prefers the recess behind the box hedge. The thrushes seem half-domesticated here; they are tame, too, in the hedges, and will sit and sing on a bough overhead without fear while you wait for a rabbit on the bank beneath.

NINE

The Orchard

B ROAD GREEN PATHS, wide enough for three or four to walk abreast, lead from the garden at Wick into the orchard. On the side next to the meadows the orchard is enclosed by a hawthorn hedge, thick from constant cropping; on the other a solid stone wall, about nine feet high, parts it from the road. One summer day a party of martins attacked this wall outside, and endeavoured to make their nest holes in it. These birds are called by the labourers 'quar martins', because they breed in holes drilled in the face of the sandy precipices of quarries. The boys 'draw' their nests – climbing up at the risk of their limbs – by inserting a long briar, and, when they feel the nest, giving it a twist which causes the hooked prickles of the stick to take firm hold, and the nest is then dragged bodily out. The flight that came to the orchard wall numbered about ten or twelve, and for the best part of the day they remained there, working their very hardest at the mortar between the stones.

The wall being old, some of the mortar had crumbled – it was not of the best quality – and here and there was a small cavity. These a portion of the birds tried to enlarge, while others boldly laboured in places where no such slight openings existed. It was interesting to watch their patient efforts as they clung to the perpendicular wall like bats. Now, two or three flew off and described a few circles in the air, as if to rest themselves, and then again returned to work. At last, convinced of the impossibility of penetrating the mortar, which was much harder beneath the surface, they went away in a body with a general twitter, leaving distinct marks of their shallow excavations. The circumstance was the more interesting because the road was much frequented (for a rural district), and many people stopped to look at them; but the birds did not seem in the least alarmed, and evidently only left because they found the wall impenetrable.

Instinct, infallible instinct, certainly would not direct these birds to such an unsuitable spot. Neither was there any peculiar advantage to attract them; it was not quiet or retired, but the reverse. The incident was clearly an experiment, and when they found it unsuccessful they desisted.

If we suppose this flight of martins to represent a party emigrating from a sand quarry (there are three such quarries within a mile radius), where the population had overflowed, it seems possible to trace the motive which animated them. I imagine that the old birds drive the young ones away, when the young return to this country with their parents after the annual migration. This is particularly the case after a very favourable breeding season, when more than the usual proportion of young birds survive. After such a season, upon returning next year to the sand quarry, the older birds drive off the younger; and if these are so numerous that they cannot find room in another part of the quarry, they emigrate in small parties.

I think the same thing happens with rooks. The older rooks will only permit a few of their last year's offspring to build near them. If a gentleman has an avenue of fine elm trees in which he desires to have a rookery, but cannot contrive to attract them, though perhaps now and then a nest is partly built and then deserted, an experiment founded on this idea might be tried. It would be necessary to ask the assistance of the proprietors of the nearest rookeries, and beg them for one year to refrain from shooting the young rooks, after the well-known custom. An unusual proportion of young birds would then survive, and next building season the larger part of these would return to the old trees and be immediately met in battle by their older relatives. Being driven away from the hereditary group of trees, they would resort to the next nearest avenue or grove; if they attempted to mix with a strange tribe, they would encounter a still fiercer resistance. In this way, the avenue in question might become stocked with rooks.

One reason, I fancy, why nests begun in such distant trees are so often deserted before completion is that a solitary nest exposes both the building birds and their prospective offspring to grave danger from hawks. No hawk will attempt to approach a rookery – the rooks would attack him *en masse* and easily put him to flight. Chickens are safer under or near a rookery from this cause: a hawk approaching them would alarm the rooks and be beaten away. The comparative safety afforded by numbers is perhaps a reason why many species of birds are gregarious. The apparently defenceless martins and swallows in this way dwell in some amount of security. If a hawk comes near the sand quarry (or the house – in the case of swallows) they all join together and pursue him, twittering angrily, and as a matter of fact generally succeed in sending him about his business. Even those birds which do not build in close contiguity no sooner find that a hawk is near than they rise simultaneously and follow and annoy him: so much so that he will sometimes actually drop the prey he has captured. It is astonishing with what temerity small birds, emboldened by numbers – chaffinches, finches, sparrows, swallows, and so on – will attack a hawk.

The quar martins that came to the orchard wall – emigrating from the quarry, and wandering about in search of a suitable habitation – if young birds, as we have supposed them to be, would naturally not yet have had much experience, and so might think the steep wall (roughly resembling the face of a quarry) available for their purpose till they had made the experiment. I have thought, from watching the motions of birds that go in flights, that most of them have a kind of leader or chief. They do not yield anything like the same obedience or reverence to the chief as the bees do to the queen bee, and exhibit little traces of following his motions implicitly. He is more like the president of a republic; each member is individually free, and twitters his or her mind just as he or she likes. But it seems to be reserved to one bird to give the signal for all to move. So these martins, after lingering about the wall for hours – some of them, too, leaving it and flying away only to come back again – finally started altogether. It is difficult to account for such simultaneous and combined movements, unless we suppose that it is reserved to a certain bird to give the signal.

In the fork of a great apple tree – a Blenheim orange – the mistle thrush has built her nest. Mistle thrushes, doubtless of the same family, have used the tree for many years. Though the nest is large, the young birds as they grow up soon get too big for it and fall out. This period – just before the young can fly – is the most critical in their existence, and causes the greatest anxiety to the parents. Without the resource of flight, weak and unable even to scramble fast through the long grass, betraying their presence by continually crying for food, they are exposed to dangers from every species of vermin.

The mistle thrush is a bold, determined bird, and does his utmost in the defence of his offspring. When the young birds fall out of the nest (so soon as one has clambered over, the others quickly follow), the parents rarely leave the orchard together. One or other is almost always close at hand. If any enemy approaches they immediately set up an angry chattering, by which noise you may at once know what is going on. I have seen two mistle thrushes attack a crow in this way. The crow came and perched upon a bough within a yard of their nest, which contained young. The old birds were there immediately, and they so annoyed and buffeted the murderous robber that he left without achieving his full purpose.

The cat is the worst enemy of the mistle thrush. It is noticeable that while these thrushes will attack anything that flies they are not so bold on the ground, but seem afraid to alight. They will strike even at the human hand that touches their nest. The crow, strong as he is, they courageously drive away; but the enemy that stealthily approaches along the ground to the helpless young bird in the grass they cannot resist. On the wing they can retreat quickly if pressed; on the ground they cannot move so swiftly, and may themselves fall a prey without affording any assistance. The mistle thrushes come to the orchard frequently after the nesting season is over and before it commences. They do not seem in search of food, but alight on the trees as if to view their property. They are strong on the wing, and fly direct to their object: there is something decided, courageous, and, as one might say, manly in their character.

The bark of some of the apple trees peels of itself – that is, the thin outer skin – and insects creep under these brown scales curled at the edges. If you sit down on the elm butt placed here as a seat and watch quietly, before long the little tree-climber will come. He flies to the trunk of the apple tree (other birds fly to the branches), and then proceeds to ascend it, going round it as he rises in a spiral. His claws cling tenaciously to the bark, his tail touches the tree, and seems to act as a support – like what I think the carpenters call a 'knee' – and his head is thrown back so as to enable him to spy into every cranny he passes. After a few turns round the trunk he is off to another tree, to resume the same restless spiral ascent there; and in a minute or so off again to a third; for he never apparently examines one half of the trunk, though, probably, his eyes, accustomed to the work, see farther than we may imagine. The orchard is never long without a tree-climber: it seems a favourite resort of these birds. They have a habit of rushing quickly a little way up; then pausing, and again creeping swiftly another foot or so, and are so absorbed in their pursuit that they are easily approached and observed.

Who can stay indoors when the goldfinches are busy among the bloom on the apple trees? A flood of sunshine falling through a roof of rosy pink and delicate white blossom overhead; underneath, grass deeply green with the vigour of spring, dotted with yellow buttercups, and strewn with bloom shaken by the wind from the trees: is not this better than formal-patterned carpets, and the white flat ceilings that weigh so heavily upon the sight? Listen how happy the goldfinches are in the orchard. Summer after summer they build in the same trees – bushy-headed codlings; generation after generation born there and gone forth to enjoy in turn the pleasures of the field.

A year – nay, a single summer – must be a very long time in their chronology, for they are so very very busy: a bright sunshiny day must be like a month to them. Now coquetting, now splashing at the sandy edge of a shallow streamlet till the golden feathers glisten from the water and the red topknot shines, away again along the hedgerow searching for seeds, singing all the while, and the tiny heart beating so rapidly as to compress twice as many beats of emotion into the minute as our sluggish organisations are capable of. Though a path much frequented by the household passes beneath the trees in which they build, they show no fear.

Just as men from various causes congregate in particular places, so there are spots in the fields – in the country generally – which appear to specially attract birds of all kinds. Wide districts are almost bare of them: on a single farm you may often find a great meadow which scarcely seems to have a bird in it, while another little oddly-cornered field is populous with them. This orchard and garden at Wick is one of the favourite places. It is like one of those Eastern marts where men of fifty different nationalities, and picturesquely clad, jostle each other in the bazaars: so here feathered travellers of every species have a kind of leafy capital. When the nesting time is over the goldfinches quit the orchard, and only return for a brief call now and then. I almost think the finches have got regular caravan routes round and across the fields which they travel in small bands.

In the meadow, just without the close-cropped hawthorn which encloses one side of the orchard, is a thick hedge, the end of which comes right up to the apple trees, being only separated by the ha-ha wall and a ditch. This hedge, dividing two meadows, is about two hundred yards long, and well grown with a variety of underwood, hazel, willow, maple, hawthorn, blackthorn, elder, and studded with some few elms and ashes, and a fine horse chestnut. Down the ditch for some distance runs a little stream (except in a long drought); and where another hedge branches from it is a hollow space arched over and roofed with boughs. Now this hedge is a favourite highway of birds and other wild creatures, and leads direct to the orchard. Most of the visitors to the house and garden come down it – it is one of their caravan routes.

If on a summer's morning you go and sit in the gateway about halfway up the hedge, partly hidden by a pollard ash and great hawthorn bushes, you will not have long to wait before you hear the pleasant calls of the greenfinches coming. They seem always to travel two or more pairs together, and constantly utter a soothing call, as if to say to their companions, 'Here we are, close by, dearest.' They all appear to know exactly where they are going – flitting across the gateway one by one, moving of one accord in the same direction; and their contented notes gradually become inaudible as they go towards the orchard. The goldfinches use the same route; so do the bullfinches. Even the starlings, before they come to the house, usually perch on an ash tree in this hedge.

There is another hedge, running parallel to it, one hundred and fifty

yards distant, the end of which also approaches the premises, but it is comparatively deserted. You may wait there in vain and see nothing but a robin.

By the same caravan route the blackbirds come to the garden; they, however, are not such travelling birds as the finches. But the tomtits are: they work their way from tree to tree for miles; they also come to the orchard by this hedge highway. As I have said before, it abuts on the orchard; and a straight line carried across to the orchard wall, over that and the road outside, would strike another great hedge which, were it not for the intervention of the garden, would be a continuation of the first. The finches, after spending a little time in the apple and damson trees, fly over the wall and road to this second hedge, and follow it down for nearly half a mile to a little enclosed meadow, which, like the orchard, is a specially favourite resort. The fondness of birds for this route is very striking; they are constantly passing up or down it. There is another such favourite route at some distance, running beside a brook and likewise leading to the same enclosed meadow – of which more presently. I think I could make a map of these fields, showing the routes and resorts of furred and feathered creatures.

Near the ha-ha wall, where the great meadow hedge comes up to the orchard, is a summer house, with a conical thatched roof and circular window. It is hung all round under the ceiling with festoons of eggs taken by the boys of the farmstead, cordially assisted by the carters' lads when not at work. There may be perhaps forty varieties, arranged so as to increase in size from the tiny tomtits up to the large wood pigeons, the peewits, corncrake, and crow: some milk-white, others splotched with dark brown spots and veins, others again blue. These eggs, when taken and the yolk blown out, were strung on a bennet and so carried home. The lads like to get them as soon after laid as possible, because they blow best then; if hard set the shell may break.

In the circular window they have left a nest of the long-tailed tit, or 'titmouse', built exactly in the shape of a nut with roof and tiny doorway, and always securely attached in the midst of a thorn bush to branches that are stiff and unlikely to bend with the breeze, so that this beautiful piece of bird-architecture may not be disturbed. To take it, it is generally necessary to cut away several boughs. Such nests are often seen in farmhouses placed as an ornament on the mantelpiece. Spiders have filled the window with their webs, and to these every now and then during the day – there is no door to the summer house – come a robin, a wren, and a flycatcher. Either of these, but more particularly the two last, will take insects from the spider's web.

The flycatcher has a favourite perch close by, and may perhaps hear the shrill buzz when an insect is caught. The flycatcher is a regular summer visitor; in the orchard, garden, and adjacent rickyard at least three pairs build every year. Under the shady apple trees near the summer house one may be seen the whole day long ever on the watch. He perches on a dead branch, low down – not up among the boughs, but as much as possible under them. Every two or three minutes he flies swiftly from his perch a few yards, darts on an insect – you cannot see it, but can distinctly hear the snap of the bill – and returns to his post. He uses the same perch for half an hour or more; then shifts to another at a little distance, and so works all round the orchard, but regularly comes back to the same spot. By waiting near it you may be certain of seeing him presently; and he is very tame, and will carry on operations within a few yards – sometimes picking up a fly almost within reach of your hand. It is noticeable that many insect-eating birds are especially tame. They will occasionally dart after a moth, but drop it again – as if they did not care for that kind of food, and yet could not resist the habit of snapping at such things.

I once saw a flycatcher rush after a buff-coloured moth, which fluttered aimlessly out of a shady recess: he snapped it, held it a second or two while hovering in the air, and then let it go. Instantly a swallow swooped down, caught the moth, and bore it thirty or forty feet high, then dropped it, when, as the moth came slowly down, another swallow seized it and carried it some yards and then left hold, and the poor creature after all went free. I have seen other instances of swallows catching good-sized moths to let them go again.

The brown linnet is another regular visitor building in the orchard; so too the blackcap, whose song, though short, is sweet; and the bold bright bullfinches use the close-cropped hawthorn. They have always a nest there, made of slender fibres dexterously interwoven. There is a group of elms near the farther end of the enclosure, and another by the rickyard; linnets seem fond of elms.

A pair of squirrels sometimes come down the same hedge – it is a favourite highway of wild animals as well as birds – to the orchard, and play in the apple trees: they even venture to a tree only a few yards from the house. If not disturbed they stay a good while, and then return by the way they came to a copse at the top of the meadow. The corner formed by the hedge and the copse – quiet, but in easy view from the house – is especially frequented by them. Their lively motions on the ground are very amusing; they visit the ground much oftener than may be generally supposed. Fir trees seem to attract them – where there is a plantation of firs you may be sure of finding a squirrel.

When alarmed or chased, a squirrel always ascends the tree on the opposite side away from you – he will not run to a solitary tree if he can possibly avoid it: he likes a group, and his trick is, the moment he thinks he is out of sight among the upper branches, to slip quietly from one tree to the other till, while you are scanning every bough, he has travelled fifty yards away unnoticed. If the branches are not close enough to hide him, he gets as much as possible behind a large branch, and stretches himself along it – at the same time his tail, which at other times is bushy, seems to contract, so that he is less visible. He will leap in his alarm to dead branches, and, though his weight is trifling, occasionally they snap under the sudden impact; but that does not distress him in the least, because a bough rarely breaks clean off, but hangs suspended by bark or splinters, so that he can scramble to the ivy that winds round the trunk. Or if he is obliged to slip down, the next branch catches him; and I have never seen a squirrel actually fall, though sometimes in his frightened haste he will send a number of little dead twigs rustling downwards. When the tail is spread out, so to say, its texture is so fine and silky that the light seems to play through it. They love this particular corner because just there the hedge is composed of hazel bushes, and even when the nuts are gone from the branches they still find some which have dropped upon the bank and are hidden in the dry grass and brown leaves.

In this corner, too, the bank being dry and sandy, there is a large settlement of rabbits, and now and then some of these find their way to the orchard and garden along the hedge. Rabbits have their own social laws and customs adapted to the special conditions of their way of life. At the breeding season there seems to be a tendency to migrate on the part of the younger rabbits from the great 'bury' hitherto their home. Many solitary holes at some distance are then occupied, and the fresh sand thrown out shows that a tenant has entered on possession. In this way one or two take up their residence more than halfway down the hedge towards the orchard. Then the doe seems to have a desire to separate herself at a certain period from the rest. She goes out into the mowing-grass perhaps thirty yards from the bury, and there the young are born in a short hole excavated for the purpose. The young rabbits naturally remain close to their birthplace; they are conducted to the hedge as soon as they are old enough to run about; and so a fresh colony is formed. As they get larger, or, say, soon after midsummer, they appear to show a tendency to roam; and by the autumn, if left undisturbed, descendants from the original settlement will have pushed outposts to a considerable distance. These, having been bred near, have little fear of entering the orchard, or even the garden, and next season will rear their offspring close at hand and feed in the enclosure, using the close-cropped hawthorn as a cover.

Weasels also occasionally come down the hedge into the orchard for the various prey they find there; they visit the outhouses and sheds, too, at intervals in the cattleyards adjoining the house. More rarely the stoat does the same. A weasel may frequently be found prowling in the highway hedge. When a weasel runs fast on a level hard surface – as across a road – the hinder quarters seem every now and then to jump up as if rebounding from the surface; his legs look too short for the speed he is going. This peculiar motion gives them when in haste an odd appearance. In a less degree, a mouse rushing in alarm across a road does the same. The motion ceases the moment mouse or weasel reaches the turf, which is rarely quite level.

The brown field mouse may be found in the orchard hedge, but is so unobtrusive that his presence is hardly observed. There are many more of these mice in the hedges than are suspected to be there; their little bodies slip about so near the surface of the brown earth, the colour of which they resemble, that few notice them unless they chance to be calling each other in their shrill treble. Even then, though the sound be audible, the mouse is invisible; but you cannot sit quiet in a hedge very long in summer without becoming aware of their presence. Some of the older branches of the hawthorn bushes, bent down when young by the hedge-cutter, are nearly horizontal and free for some part of their length of twigs. The mice run along these natural bridges from one part of the hedge to the other.

Last spring I watched a mouse very busily engaged sitting on such a branch, about a foot above the bank, nibbling the tender top leaves of the 'clite' plant. The clite grows with great rapidity, and climbs up into the hedge; this plant had already pushed up ten or twelve inches, so that the mouse on the branch was just about on a level with the upper and tenderest leaves. These he drew towards him with his fore feet and complacently nibbled. When he had picked out what suited his fancy he ran along the branch, and in an instant was lost to sight on the bank among the grass.

The nests of the 'harvest trow' – a still smaller mouse, seldom seen except in summer – are common in the grass of the orchard (and in almost every meadow) before it is mown. As the summer wanes their dead bodies are frequently found in the footpaths; for a kind of epizoic seems to seize them at that time, and they die in numbers. It is curious that an animal which carefully conceals itself in health should at the approach of death seek an open and exposed place like a footpath worn clear of grass.

In the ha-ha wall, at that part of the orchard where the highway hedge comes up, is the square mouth of a rather large drain. The drain itself is of rude construction – two stones on edge and a third across at the top. It comes from the cowyard, passing under the outermost part of the garden a considerable distance away from the house. Very early one morning the labourers, coming to work, saw a fox slip into the mouth of the drain through the long grass of the meadow on which it opened. In the summer, the cattle being all out in the fields, the drain was perfectly dry, and it was known that now and then the rabbits from the hedge made use of it as a temporary place of concealment. No doubt the presence of a rabbit in it was the cause of the fox entering in the first place. The rabbit must have had a very bad time of it, for, the drain being closed at the other end with an iron grating, no possibility of escape existed.

From the traces in the grass and on the dry mud at the mouth it appeared as if the fox had ventured there more than once; and, as there were many chickens about, his object in lying here was evident. The great hedge being so near, and the narrow space between full of tall mowing-grass – the edge of the ha-ha wall, too, clothed with stonecrop and grasses growing in the interstices of the loose stones, and further sheltered by a low box hedge – it was a place almost made on purpose for Reynard's cunning ambuscade. He is as bold or even bolder than he is cunning. A young dog sent up the drain came back quicker than he went, and refused to venture a second time. The fox remained there all day, and of course made tracks at night, knowing that his presence had been discovered by the commotion and talking at the mouth of his cave. He might easily have been captured, but that was not attempted on account of the hunt.

Though the fox as a general rule sleeps during the day, he does not always, but sometimes makes a successful foray in broad daylight. Fowls, for instance, at night roost in the sheds at some height from the ground – often the sheds are contrived specially to protect them; but in the day they roam about in the vicinity of the rickyards where they are kept. They will make runs down the centre of a double-mound hedge, and while thus rambling occasionally stroll into the jaws of their foe, who has been patiently waiting hidden in the long grass and underwood. In the day, too, rabbits often sit out in a bunch of grass, or dry furrow, a long way from the bury. Their form is usually within a few paces of a well-marked run – they follow the run out into the field, and then leave it and go among the grass at one side. The run, therefore, sometimes acts as a guide to the fox, who, sheltered by the tall bennets and thick bunches, occasionally glides up it in the daytime to his prey.

There is sure to be a snake or two in the grass of the orchard during the summer, especially if there chance to be an old manure heap anywhere near; for that is the place in which they like to leave their chains of white eggs, out of which, if broken, the little snakes issue only two or three inches in length. The heat of the manure heap acts as an incubator.

When it is wet and the hay cannot be touched, the haymakers, there being nothing else for them to do, are put to turn such heaps, and frequently find the eggs of snakes. These creatures now and then get inside farmhouses, whose floors are generally on a level with the surface of the earth or nearly so. They have been found in the clock case – the old upright eight-day clock, standing on the floor; they come after the frogs that enter at the doors – always wide open in summer – and are supposed also to eat crumbs.

In the cellar there is sure to be a toad under the barrels on the cool stone flags; in the garden there is another, purposely kept in the cucumber frame to protect the plant from being eaten by creeping things. It is curious to notice that they both seem to flourish equally well – one in the coolest, the other in the hottest place. A third may generally be found in the strawberry bed. Strawberries are much eaten by insects of many kinds; so that the toad really does good service in a garden.

In winter, when snow is on the ground, a few larks sometimes venture into the garden where anything green yet shows above the white covering on the patches. If the weather is severe, the moorhen will come up from the brook, though two fields distant, in the night, and the marks of her feet may be traced round the house. Then, as the evening approaches, the wild ducks pass over, and every now and then during the night the weird cries of waterfowl resound in the frosty air. The heron sails slowly over, every night and every morning, backwards and forwards from the mere to the water meadows and the brook, uttering his unearthly call at intervals.

TEN

The Woodpile

THE FARMHOUSE AT WICK has the gardens and orchard upon one side, and on the other are the cart houses, sheds, and rickyards. Between these and the dwelling runs a broad roadway for the waggons to enter and leave the fields, and on its border stands a great woodpile. The faggots cut in the winter from the hedges are here stacked up as high as the roof of a cottage, and nearby lies a heap of ponderous logs waiting to be split for firewood. From exposure to the weather the bark of the faggot sticks has turned black and is rapidly decaying, and under it innumerable insects have made their homes.

For these, probably, the wrens visit the woodpile continually; if in passing any one strikes the faggots with a stick, a wren will generally fly out on the opposite side. They creep like mice in between the faggots – there are numerous interstices – and thus sometimes pass right through a corner of the stack. Sometimes a pole which has been lying by for a length of time is found to be curiously chased, as it were, all over the surface under the loose bark by creeping things. They eat channels interweaving and winding in and out in an intricate pattern, occasionally a little resembling the Moorish style of ornamentation seen on the walls of the Alhambra. I have found poles so curiously carved like this that the idea naturally occurred of using them for cabinet work. They might at least have supplied a hint for a design. Besides the wrens, many other birds visit the woodpile – sparrows are perpetually coming, and on the retired side towards the meadow the robins build their nests. On the ridge where some of the sticks project the swallows often perch and twitter – generally a pair seem to come together.

It takes skill as well as mere strength even to do so simple a thing as to split the rough logs lying here on the ground. They are not like those

Abraham Lincoln began life working at – even-grained wood, quickly divided – but tough and full of knots strangely twisted; so that it needs judgment to put the wedges in the right place.

Near the woodpile is a well and a stone trough for thirsty horses to drink from, and as the water, carelessly pumped in by the carters' lads, frequently overflows, the ground just there is usually moist. If one of the loose oak logs that lie here with the grass growing up round it is rolled over, occasionally a lizard may be found under it. This lizard is slender, and not more than three or four inches in length, general colour a yellowish green. Where one is found a second is commonly close by. They are elegantly shaped, and quick in their motions, speedily making off. They may now and then be discovered under large stones, if there is a crevice, in the meadows. They do not in the least resemble the ordinary land lizard, which is a much coarser-looking and larger creature, and it is not an inhabitant of this locality: at all events it is rare enough to have escaped me here, though I have often observed it in districts where the soil is light and sandy and where there is a good deal of heathland. The land lizards will stroll indoors if the door be left open. These lesser but more elegant lizards appear to prefer a damp spot – cool and moist, but not positively wet.

A large shed built against the side of the adjacent stable is used as a carpenter's workshop – much carpenter's work is done on a farm – and here is a bench with a vice and variety of tools. When sawing, the wood operated on often 'ties' the saw, as it is called – that is, pinches it – which makes it hard to work; a thin wedge of wood is then inserted to open a way, and the blade of the saw rubbed with a little grease, which the metal, heated by the friction, melts into oil. This eases the work – a little grease, too, will make a gimlet bore quicker. Country carpenters keep this grease in a horn – a cow's horn stopped at the larger end with a piece of wood and at the other by its own natural growth. Now the mice (which are everywhere on farm premises) are so outrageously fond of grease that they will spend any length of time gnaw-gnaw-gnawing till they do get at it. Right through the solid stopper of wood they eat their way, and even through the horn; so that the carpenter is puzzled to know how to preserve it out of their reach.

It is of no use putting it on a shelf, because they either rush up the wall or drop from above. At last, however, he has hit upon a dodge.

He has suspended the horn high above the ground by a loop of copper wire, which projects six or eight inches from the wall, like a lamp on a bracket. The mice may get on the bench, and may run up the wall, but when they get to the wire they cannot walk out on it – like tightrope walking – the more especially as the wire, being thin and flexible, bends and sways if they attempt it. This answers the purpose as a rule; but even here the carpenter declares that once now and then his horn is pilfered, and can only account for it by supposing that a bolder mouse than common makes a desperate leap for it, and succeeds in landing on the flat surface of the wooden stopper.

The shed has one small window only, which has no glass, but is secured by an iron bar (he needs no larger window, for all carpenters work with the door open); and through this window a robin has entered and built a nest in a quiet corner behind some timber. Though a man is at work here so often, hammering and sawing, the birds come fearlessly to their young, and pick up the crumbs he leaves from his luncheon.

Between the timber framework of the shed and the brickwork of the adjacent stable chinks have opened, and in these and in the chinks between the wooden lintel of the stable door and the bricks above it the bats frequently hide, passing the day there. Others hide in the tiles of the roof where their nests are made. The labouring lads often amuse themselves searching for these creatures, whose one object in daylight seems to be to cling to something; they will hang to the coat with the claws at the extremity of their membranous wings, and if left alone will creep out of sight into the pocket. There are two well-marked species of bats here – one small and the other much larger.

The lesser bat flies nearer to the ground, and almost always follows the contour of some object or building. They hawk to and fro for hours in the evening under the eaves of the farmhouse, and frequently enter the great garrets and the still larger cheese room (where the cheese is stored to mature) – sometimes through the windows, and sometimes seeming to creep through holes made by sparrows or starlings in the roof. Moths are probably the attraction; of these there are generally plenty in and about old houses. Occasionally a bat will come into the sitting room, should the doors be left open on a warm summer evening: this the old folk think an evil omen, and still worse if in its alarm at the attempts made to drive it away it should chance to knock against a candle and overturn or put it out. They think, too, that a bat seen in daytime is a bad sign. Once now and then one gets disturbed by some means in the tiles, and flutters in a helpless manner to the nearest shelter; for in daylight they seem quite at a loss, though flying so swiftly at night.

The greater bat hawks at a considerable elevation above houses and trees, and wheels and turns with singular abruptness, so that some think it a test of a good shot to bring them down. The reason, however, why many find it difficult to hit a bat is because they are unaccustomed to shooting at night, and not because of its manner of flight, for it often goes quite straight. It is also believed to be a test of good hearing to be able to hear the low shrill squeak of the bat uttered as it flies: the same is said of the shrew, whose cry is yet more faint and acute. The swift, too, has a peculiar kind of screech, but easily heard.

Beyond the stables are the cattle sheds and cowyards. These sheds are open on the side towards the yard, supported there by a row of wooden pillars stepped on stones to keep them from rotting. On the large crossbeams within the swallows make their nests. When the eggs are hard set, the bird will sit so close that with care and a gentle manner of approach you may sometimes even stroke her back lightly with your finger without making her rise. They become so accustomed to men constantly in and out the sheds as to feel little alarm. Some build their nests higher up under the roof tree.

To the adjoining rickyard redstarts come every summer, building their nests there; 'horse matchers' or stonechats also in summer often visit the rickyard, though they do not build in it. Some elm trees shade the ricks, and once now and then a wood pigeon settles in them for a little while. The coo of the dove may be heard frequently, but she does not build very near the house.

On this farm the rookery is at some distance in the meadows, and the rooks rarely come nearer than the field just outside the post and rails that enclose the rickyard, though they pass over constantly, flying low down without fear, unless someone chances just then to come out carrying a gun. Then they seem seized with an uncontrollable panic, and stop short in their career by a violent effort of the wings, to wheel off immediately at a tangent. Perhaps no other bird shows such evident signs of recognising a gun. Chaffinches, it must not be forgotten, frequent the rickyard in numbers.

Finally come the rats. Though trapped, shot, and ferreted without mercy, the rats insist on a share of the good things going. They especially haunt the pigsties, and when the pigs are served with their food feed with them at the same trough. Those old rats that come to the farmstead are cunning, fierce beasts, not to be destroyed without much difficulty. They will not step on a trap, though never so cleverly laid; they will face a ferret, unless he happens to be particularly large and determined, and bite viciously at dogs. But with all their cunning there is one simple trick which they are not up to: this is to post yourself high up above the ground, when they will not suspect your presence; a ladder is placed against a tree within easy shot of the pigsty, and the gunner, having previously arranged that everything shall be kept quiet, takes his stand on it, and from thence kills a couple perhaps at once.

On looking back, it appears that the farmhouse, garden, orchard, and rickyard at Wick are constantly visited by about thirty-five wild creatures, and, in addition, five others come now and then, making a total of forty. Of these forty, twenty-six are birds, two bats, eight quadrupeds, and four reptiles. This does not include some few additional birds that only come at long intervals, nor those that simply fly overhead or are heard singing at a distance.

The great meadow hedge – the highway of the birds – where it approaches the ha-ha wall of the orchard, is lovely in June with the wild roses blooming on the briars which there grow in profusion. Some of these briars stretch forth into the meadow, and then, bent down by their own weight, form an arch crowned with flowers. There is an old superstition about those arches

of briar hung out along the hedgerow: magical cures of whooping cough
and some other diseases of childhood can, it is believed, be effected by
passing the child at sunrise under the briar facing the rising sun.

This had to be performed by the 'wise woman'. There was one in every
hamlet until a few years ago – and indeed here and there an aged woman
retains something like a reputation for witchcraft still. The wise woman
conducted the child entrusted to her care at the dawn to the hedge, where
she knew there was a briar growing in such a position that a person could
creep under it facing the east, and there, as the sun rose, passed the child
through.

In the hollow just beneath the ha-ha wall, where it is moist, grow tall
rushes; and here the great dragonfly darts to and fro so swiftly as to leave
the impression of a line of green drawn suddenly through the air. Though
travelling at such speed, he has the power of stopping abruptly, and
instantly afterwards returns upon his path. These handsome insects are
often placed on mirrors as an ornament in farmhouses. The labourers will
have it that they sting like the hornet; but this they say also of many other
harmless creatures, seeming to have a general distrust of the insect kind.
They will tell you alarming stories of terrible sufferings – arms swollen to
double the natural size, necks inflamed, and so forth – caused by the bites
of unknown flies. Not being able to discover what fly it is that inflicts these
poisonous wounds, and having spent so many hours in the fields without
experiencing such effects, I rather doubt these statements, though put forth
in perfect good faith: indeed, I have often seen the arms and chests of the
men in harvest time with huge bumps rising on them which they declared
were thus caused. The common harvest bug, which gets under the skin,
certainly does not cause such great swellings as I have seen; nor the stoatfly,
which latter is the most bloodthirsty wretch imaginable.

With a low, hissing buzz, a long, narrow, and brownish-grey insect settles
on your hand as you walk among the hay, and presently you feel a tingling
sensation, and may watch (if you have the patience to endure the irritation)
its body gradually dilate and grow darker in colour as it absorbs the blood.
When once thoroughly engaged, nothing will frighten this fly away: you

may crush him, but he will not move from fear: he will remain till, replete with blood, he falls off helpless into the grass.

The horses in the waggons have at this season to be watched by a boy armed with a spray of ash, with which he flicks off the stoats that would otherwise drive the animals frantic. A green spray is a great protection against flies; if you carry a bough in your hand as you walk among the meadows they will not annoy you half as much. Such a bough is very necessary when lying *perdu* in a dry ditch in summer to shoot a young rabbit, and when it is essential to keep quiet and still. Without it it is difficult to avoid lifting the hand to knock the flies away – which motion is sure to alarm the rabbit that may at that very moment be peeping out preparatory to issuing from his hole. It is impossible not to pity the horses in the hayfields on a sultry day; despite all care taken, their nostrils are literally black with crowds of flies, which constantly endeavour to crawl over the eyeball. Sunshine itself does not appear so potent in bringing forth insects as the close electrical kind of heat that precedes a thunderstorm. This is so well known that when the flies are more than usually busy the farmer makes haste to get in his hay, and lets down the canvas over his rick. The cows give warning at the same time by scampering about in the wildest and most ludicrous manner – their tails held up in the air – tormented by insects.

The ha-ha wall, built of loose stones, is the home of thousands upon thousands of ants, whose nests are everywhere here, the ground being undisturbed by passing footsteps. They ascend trees to a great height, and may be seen going up the trunk sometimes in a continuous stream, one behind the other in Indian file.

In one spot on the hedge of the ha-ha is a row of beehives – the garden wall and a shrubbery shelter them here from the north and east, and the drop of the ha-ha gives them a clear exit and entrance. This is thought a great advantage – not to have any hedge or bush in front of the hives – because the bees, heavily laden with honey or pollen, encounter no obstruction in coming home. They are believed to work more energetically when this is the case, and they certainly do seem to exhibit signs of annoyance, as if out of temper, if they get entangled in a bush. Indeed, if you chance

to be pursued by an angry cloud of bees whose ire you have aroused, the only safe place is a hedge or bush, into which make haste to thrust yourself, when the boughs and leaves will baffle them. If the hive be moved to a different place, the bees that chance at the time to be out in the fields collecting honey, upon their return, finding their home gone, are evidently at a loss. They fly around, hovering about over the spot for a long time before they discover the fresh position of the hive.

The great hornet, with its tinge of reddish orange, comes through the garden sometimes with a heavy buzz, distinguishable in a moment from the sound of any other insect. All country folk believe the hornet's sting to be the most poisonous and painful of any, and will relate instances of persons losing the use of their arms for a few days in consequence of the violent inflammation. Sometimes the hornet selects for its nest an aperture in an old shed near the farmhouse. I have seen their nests quite close to houses; but, unless wantonly disturbed, there is not the slightest danger from them, or indeed from any other insect of this class. I think the common hive-bees are the worst tempered of any – they resent the slightest interference with their motions. The hornet often chooses an old hollow withy-pollard for the site of its nest.

In the orchard there is at least one nest of the humblebee, made at a great depth in a deserted mouse's hole. These bees have eaten away and removed the grass just round the entrance, so as to get a clear road in and out. They are as industrious as the hive-bee; but, as there are not nearly so many working together in one colony, they do not store up anything approaching to the same quantity of honey. There is a superstition that if a humblebee buzzes in at the window of the sitting room it is a sure sign of a coming visitor.

Be careful how you pick up a ripe apple, all glowing orange, from the grass in the orchard; roll it over with your foot first, or you may chance to find that you have got a handful of wasps. They eat away the interior of the fruit, leaving little but the rind, and this very hollowness causes the rind to assume richer tints and a more tempting appearance. Speckled apples on the tree, whether pecked by a blackbird, eaten by wasps or ants, always ripen fastest, and if you do not mind cutting out that portion, are the best. Such a fallen apple, when hollowed out within, is a veritable torpedo if incautiously handled.

Wasps are incurable drunkards. If they find something sweet and tempting they stick to it, and swill till they fall senseless to the ground. They are then most dangerous, because unseen and unheard; and one may put one's hand on them in ignorance of their whereabouts. Noticing once that a particular pear tree appeared to attract wasps, though there was little or no fruit on it, I watched their motions, and found they settled at the mouth of certain circular apertures that had been made in the trunk. There the sap was slowly exuding, and to this sap the wasps came and sipped it till they could sip no more. The tree being old and of small value, it was determined to see what caused these circular holes. They were cut out with a gouge, when the whole interior of the trunk was found bored with winding tunnels, through which a pistol bullet might have been passed. This had been done by an enormous grub, as long and large as one's finger.

Old world plants and flowers linger still like heirlooms in the farmhouse garden, though their pleasant odour is oft-times choked by the gaseous fumes from the furnaces of the steam-ploughing engines as they pass along the road to their labour. Then a dark vapour rises above the tops of the green elms, and the old walls tremble and the earth itself quakes beneath the pressure of the iron giant, while the atmosphere is tainted with the smell of cotton waste and oil. How little these accord with the quiet, sunny slumber of the homestead. But the breeze comes, and ere the rattle of the wheels and cogs has died away, the fragrance of the flowers and green things has reasserted itself. Such a sunny slumber, and such a fragrance of flowers, both wild and cultivated, have dwelt round and over the place these two hundred years, and mayhap before that. It is perhaps a fancy only, yet I think that where men and nature have dwelt side by side time out of mind there is a sense of a presence, a genius of the spot, a haunting sweetness and loveliness not elsewhere to be found. The most lavish expenditure, even when guided by true taste, cannot produce this feeling about a modern dwelling.

At Wick, by the side of the garden path, grows a perfect little hedge of lavender; every drawer in the house, when opened, emits an odour of its dried flowers. Here, too, are sweet marjoram, rosemary, and rue; so also bay and thyme, and some pot-herbs whose use is forgotten, besides southernwood and wormwood. They do not make medical potions at home here now, but the lily leaves are used to allay inflammation of the skin. The house leek had a reputation with the cottage herbalists; it is still talked of, but I think very rarely used.

Among the flowers here are beautiful dark-petalled wallflowers, sweet williams, sweet briar, and pansies. In spring the yellow crocus lifts its head from among the grass of the green in front of the house (as the snowdrops did also), and here and there a daffodil. These, I think, never look so lovely as when rising from the greensward; the daffodils grow, too, in the orchard. Woodbine is everywhere – climbing over the garden seat under the sycamore tree, whose leaves are spotted sometimes with tiny reddish dots, the honeydew.

Just outside the rickyard, where the grass of the meadow has not been mown but fed on by cattle, grow the tall buttercups, rising to the knee. The children use the long hollow stems as tubes wherewith to suck up the warm new milk through its crown of thick froth from the oaken milking pail. There is a fable that the buttercups make the butter yellow when they come – but the cows never eat them, being so bitter; they eat all round close up to the very stems, but leave them standing scrupulously. The children, too, make similar pipes of straw to suck up the new cider fresh from the cider mill, as it stands in the tubs directly after the grinding. Under the shady trees of the orchard the hare's parsley flourishes, and immediately without the orchard edge, on the shore of the ditch, grow thick bunches of the beautiful blue cranesbill, or wild geranium, which ought to be a garden flower and not left to the chance mercy of the scythe. There, too, the herb Robert hides, and its foliage, turning colour, lies like crimson lace on the bank.

Even the tall thistles of the ditch have their beauty – the flower has a delicate tint, varying with the species from mauve to purple; the humblebee visits every thistle-bloom in his path, and there must therefore be sweetness in it. Then in the autumn issues forth the floating thistledown, streaming through the air and rolling like an aerial ball over the tips of the bennets. Thistledown is sometimes gathered to fill pillowcases, and a pillow so filled is exquisitely soft. There is not a nook or corner of the old place where something interesting may not be found. Even the slates on a modern addition to the homestead are each bordered with yellow lichen – perhaps because they adjoin thatch, for slates do not seem generally to encourage the growth of lichen. It appears to prefer tiles, which therefore sooner assume an antique tint.

To the geraniums in the bow window the hummingbird moth comes now and then, hovering over the scarlet petals. Out of the high elms drops a huge grey moth, so exactly the colour of grey lichen that it might be passed for it – pursued, of course, as it clumsily falls, by two or more birds eager for the spoil. It is feast time with them when the cockchafers come: they leave nothing but wing cases scattered on the garden paths like the shields of slain men-at-arms.

In the bright sunshine, when there is not a cloud in the sky, slender beetles come forth from the cracks of the earth and run swiftly across the paths, glittering green and gold, iridescent colours glistening on their backs. These are locally called 'sun beetles', because they appear when the sun is brightest. Be careful not to step on or kill one; for if you do it will certainly rain, according to the old superstition. The blackbird, when he picks up one of the larger beetles, holds it with its back towards him in his bill, so that the legs claw helplessly at the air, and thus carries it to a spot where he can pick it to pieces at his leisure.

The ha-ha wall of the orchard is the favourite haunt of butterflies; they seem to love its sunny aspect, and often cling to the loose stones like ornaments attached by some cunning artist. Sulphur butterflies hover here early in the spring, and later on white and brown and tiny blue butterflies pass this way, calling *en route*. Sometimes a great noble of the butterfly world comes in all the glory of his wide velvety wings, and deigns to pause a while that his beauty may be seen.

Somewhere within doors, in the huge beams or woodwork, the deathtick is sure to be heard in the silence of the night; even now the old folk listen with a lingering dread. Give the woodwork a smart tap, and the insect stops a few moments, but it rapidly gets accustomed to such taps, and after a few, ceases to take notice of them. This manner of building houses with great beams visibly supporting the ceiling, passing across the room underneath it, had one advantage. On a rainy day the children would go into the garrets or the cheese loft and there form a swing, attaching the ropes to the hooks in the beam across the ceiling.

The brewhouse, humble though its object may be, is not without its claim to admiration. It is open from the floor to the rafters of the roof, and that roof in its pitch, the craft of the woodwork, the dull polish of the old oak, has an interest far surpassing the dead staring level of flat lath and plaster. Noble workmanship in wood may be found, too, in some of the ancient barns; sometimes the beams are of black oak, in others of chestnut.

In these modern days men have lost the pleasures of the orchard; yet an old-fashioned orchard is the most delicious of places to idle away the afternoon of a hazy autumn day, when the sun seems to shine with a soft slumbrous warmth without glare, as if the rays came through an aerial spider's web spun across the sky, letting all the beauty, but not the heat, slip through its invisible meshes. There is a shadowy coolness in the recesses under the trees. On the damson trunks are yellowish crystalline knobs of gum which has exuded from the bark. Now and then a leaf rustles to the ground, and at longer intervals an apple falls with a decided thump. It is silent save for the gentle twittering of the swallows on the topmost branches – they are talking of their coming journey – and perhaps occasionally the distant echo of a shot where the lead has gone whistling among a covey. It is a place to dream in, bringing with you a chair to sit on – for it will be freer from insects than the garden seat – and a book. Put away all thought of time: often in striving to get the most value from our time it slips from us as the reality did from the dog that greedily grasped at the shadow: simply dream of what you will with apples and plums, nuts and filberts within reach.

Dusky Blenheim oranges, with a gleam of gold under the rind; a warmer tint of yellow on the pippins. Here streaks of red, here a tawny hue. Yonder a load of great russets; nearby heavy pears bending the strong branches; round black damsons; luscious egg plums hanging their yellow ovals overhead; bullace, not yet ripe, but presently sweetly piquant. On the walnut trees bunches of round green balls – note those that show a dark spot or streak, and gently tap them with the tip of the tall slender pole placed there for the purpose. Down they come glancing from bough to bough, and, striking the hard turf, the thick green rind splits asunder, and the walnut itself rebounds upwards. Those who buy walnuts have no idea of the fine taste of the fruit thus gathered direct from the tree, when the kernel, though so curiously convoluted, slips its pale yellow skin easily and is so wondrously white. Surely it is an error to banish the orchard and the fruit garden from the pleasure-grounds of modern houses, strictly relegating them to the rear, as if something to be ashamed of.

ELEVEN

The Homefield

A WICKET GATE affords a private entrance from the orchard into the homefield, opening on the meadow close to the great hedge, the favourite highway of the birds. Tracing this hedge away from the homestead, in somewhat more than two hundred yards it is joined by another hedge crossing the top of the field, thus forming a sheltered nook or angle, which has been alluded to as the haunt of squirrels. Here the highway hedge is almost all of hazel, though one large hawthorn tree stands on the shore of the ditch. Hazel grows tall, straight, and is not so bushy as some underwood; the lesser boughs do not interlace or make convenient platforms on which to build nests, and birds do not use it much.

The ancient divination by the hazel wand, or, rather, the method of searching for subterranean springs, is not yet forgotten; some of the old folk believe in it still. I have seen it tried myself, half in joke, half in earnest. A slender rod is cut, and so trimmed as to have a small fork at one end; this fork is placed under the little finger in such a way that the rod comes over the back of the other fingers; it is then lightly balanced, and vibrates easily. The magician walks slowly over the ground selected, watching the tip of the wand; and should it bend downwards without volition on his part, it is a sign that water is concealed beneath the spot.

The nuts upon the bushes do not all ripen at the same time: one or two bushes are first, and offer ripe nuts before the rest have hardened sufficiently. The leaves on these also drop earlier, turning a light yellow. The size and even the shape of the nuts vary too, some being nearly round and others roughly resembling the almond. Their flavour when taken from the bush is sweet, juicy, nutty. When they will 'slip udd' is the proper time to gather them – i.e. when the hood or outer green covering slips off at a touch, leaving the light-brown nut in the palm: it is a delicately shaded brown. Cut off just the tip of the nut – the pointed keystone of its Gothic arch – with a penknife; insert the blade ever so slightly, and a gentle turn splits the shell and shows two onyx-white hemispheres of kernel.

With a little care the tallest boughs may be pulled down uninjured; if dragged down rudely the bough will be sprung where it joins the stole below, and will then wither and die. The plan is simply to apply force by degrees, pulling the main bough only so far forward as to enable the hand to reach an upper branch, seizing the upper branch, and by its aid reaching a still higher one, and gradually bending the central stem till it forms a bow. If done gradually and the bow not too acute, the tallest bush will spring up when released without the least injury. With a crook to seize the bush as high up as possible – where it bends more easily – not a twig need be broken, and nutting may be enjoyed without doing the least damage.

Under a tall ash tree rising out of the hazel bushes, and near the great hawthorn on the edge or shore of the ditch, the grass grows rank and is of the deepest green. The dove that could be heard cooing from the orchard built her nest in the hawthorn, which, where it overhangs the grass like a canopy, is bare of boughs for six or seven feet up the gnarled stem. The cattle, who love to shelter under it from the heat of the sun, browsed on the young shoots, so that no branch could form; but on the side towards the ditch there are immense spiny thorns, long enough and strong enough to make a savage's arrowhead or awl. The doves do not seem nearly so numerous as the wood pigeons (doves, too, in strict language); they are much smaller, rather duller in colour – that is, when flying past – and are rarely seen more than two together. When the summer thunder is booming yonder over the hills, and the thin edge of the dark cloud showers its sweet refreshing rain, with the sunshine gleaming through on the hedge and grass here, between the rolling echoes the dove may be heard in the bush coo-cooing still more softly and lovingly to her mate.

Just in the very angle formed by the meeting hedges the ditch becomes almost a fosse, so broad and deep; the sandy banks have slipped, and the rabbits have excavated more, and over all the brambles have arched thickly with a background of brake fern. The flower of the bramble is very beautiful – a delicate pink bloom, succeeded by green berries, to ripen red, and later black, under the sun. A larger kind are found here and there – the children call them dewberries. Some of the bramble leaves linger on a dull green all through the winter.

In the angle a narrow opening runs through between the two banks, which do not quite meet: it is so overgrown with bramble and fern, convolvulus and thorn, that unless the bushes were parted to look in no one would suspect the existence of this green tunnel, which on the other side opens on the ash copse, where a shallow furrow (dry) joins it. This tunnel is the favourite way of the rabbits from the copse out into the tempting pasturage of the meadow; through it, too, now and then, a fox creeps quietly. Rabbit holes drill the bank everywhere, but one near this green byway is noticeable because of its immense size.

It must measure eighteen inches in diameter at the mouth; nor does it diminish abruptly, but continues almost as large a yard or more inside the bank. Spaniels will get right into such a bury, till nothing but the tail can be seen, and, if permitted, stay there and dig and scratch frantically. They would sometimes, perhaps, succeed in reaching the prey were it not for the roots of thorn bushes or trees which cross the holes here and there like bars; these they cannot scratch through, but will bite and tear with their teeth – coming out now and then to breathe and shake the sand from their muzzles, then back again with a whine of eager excitement, till presently, in sheer exhaustion, they lie down at the mouth of the cave and pant. This is not allowed if it is known; but spaniels now and then steal away privately, and so frequently make for a hole like this that when their absence is discovered it is the first place visited in search of them. The mingled patience and excitement, the vast labour they will undergo, the quantity of sand they will throw out, the whine – it is not a bark – expressing intense desire, prove how deep is the hunting instinct in the dog.

Even if the burrows be ferreted, in a few weeks this great hole shows signs of fresh inhabitants; and such a specially enlarged entrance may be found somewhere in most of the banks frequented by rabbits. Why do they make an aperture so many times larger than they can possibly require? It may be a kind of ancestral hall, the favourite cave of the first settlers here, clung to by their descendants. Within, perhaps three, or even more, tunnels branch off from it. So busy are they, and so occupied when excavating a fresh passage, that sometimes when waiting quietly on a bank you may see the miner at work. The sand pours out as he casts it behind him with his hinder paws; his back is turned, so that he does not notice anyone.

Along the banks evidence may be found of attempts at boring holes, abandoned after a few inches of progress had been made; sometimes a root, or a stone perhaps, interferes; sometimes, and apparently more often, caprice seems the only cause why the tunnel was discontinued. The grass in this corner is sweeter to their taste than elsewhere: their runs are everywhere – crossing and winding about.

In the evening, as the shadows deepen and a hush falls upon the meads, they come out and chase and romp with each other. When a couple are at play one will rush ten or a dozen yards away and begin to nibble as if totally unconscious of the other. The second meanwhile nibbles too, but all the while stealthily moves forward, not direct, but sideways, towards the first, demurely feeding. Suddenly the second makes a spring; the first, who has been watching out of the corners of his eye all the time, is off like the wind. Or sometimes he will turn and face the other, and jump clean over, a foot high. Sometimes both leap up together in the exuberance of their mirth.

By the trunk of a mighty oak, growing out of the hedge that runs along the top of the field, the brambles and underwood are thinner, as is generally the case close under a tree; and it is easy to push through just there. On the other side, a huge root covered with deep green moss affords a pleasant seat, leaning back against the trunk. Upon the right, close by, is the ash copse, with its border of thick fir trees; on the left oaks at intervals stand along the hedge; in front stretches the undulating surface of an immense pasture field called The Warren. Like a prairie it rolls gently away, dotted with hawthorn bushes, here and there a crab tree, and two rows of noble elms, in both of which the rooks are busy in spring. Beyond, the ground rises, and the small upland meadows are so thickly timbered as to look like distant glades of a forest; still farther are the downs.

Under this great oak in the stillness is a place to dream – in summer, looking upward into the vast expanse of green boughs is an intricate architecture, an inimitable roof, whose lattice windows are set with translucent *lapis lazuli*, for the deep blue of the sky seems to come down and rest upon it. The acorns are already there, as yet all cup, and little of the acorn proper showing; there is a tiny black speck on the top, and the young acorn faintly resembles some of the ancient cups with covers, the black speck being the knob by which the cover is lifted. After the first frosts, when the acorns are browned and come out of their cups from their own weight as they fall and strike the ground, the lads select the darkest or ripest, and eat one now and then; they half-roast them too, like chestnuts.

In the early spring, when the night is bright and clear, it is a place to stand a moment and muse awhile. For the copse is dark and gloomy, the bare oaks are dark behind; the eye cannot see across the prairie, whose breadth is doubled by the night. But yonder lies a great grey sarsen boulder, like an uncouth beast of ancient days crouching in the hollow. Hush! there was a slight rustling in the grass there, as of a frightened thing; it was a startled hare hastening away. The brightest constellations of our latitude pour down their rays and influence on the birth of bud and leaf in spring; and at no other season is the sky so gorgeous with stars.

The grass in homefield as it begins to grow tall in spring is soon visited by the corncrakes, who take up their residence there. In this district (though called the corncrake) these birds seem to frequent the mowing-grass more than the arable fields, and they generally arrive about the time when it has grown sufficiently high and thick to hide their motions. This desire of concealment is apparently more strongly marked in them than in any other bird; yet they utter their loud call of 'Crake, crake, crake!' not unlike the turning of a wooden rattle, continuously though only at a short distance.

It is difficult to tell from what place the cry proceeds: at one moment it sounds almost close at hand, the next fifty yards off; then, after a brief silence, a long way to one side or the other. The attempt to mark the spot is in vain; you think you have it, and rush there, but nothing is to be seen, and a minute afterwards 'Crake, crake!' comes behind you. For the first two or three such attempts the crake seems to move but a little way, dodging to and fro in a zigzag, so that his call is never very far off; but if repeated again and again he gets alarmed, there is a silence, and presently you hear him in a corner of the mead a hundred yards distant. Perhaps once, if you steal up very, very quietly, and suddenly dart forward, or if you have been waiting till he has come unawares close to you, you may possibly see the grass move as if something passed through it; but in a moment he is gone, without a glimpse of his body having been seen. His speed must be very great to slip like this from one side of the field to the other in so few seconds.

The fact that the call apparently issues from the grass in one place, and yet upon reaching it the bird is not to be found, has given rise to the belief that the crake is a ventriloquist. It may be so; but, even without special powers of that kind, ventriloquial effects would, I think, be produced by the peculiar habits of the bird. When that which causes a sound is out of sight it must always be difficult to fix upon the exact spot whence the sound comes. When the sound is made now here, now yonder, as the bird travels swiftly – still out of sight – it must be still more difficult. The crake doubtless often cries from a furrow which would act something like a trough, tending to draw the sound along it. Finally, the incessant repetition of the same note, harsh and loud, confuses the ear.

Some say in like manner that the starling ventriloquises. He has, indeed, one peculiar long-drawn hollow whistle which goes echoing round the chimney pots and to and fro among the gables; but it never deceives you as to his position on the roof unless you are indoors and cannot see him. It is the same with the finches in the trees, when the foliage is thick. Their notes seem to come from this side among the branches, but on peering carefully up there is no bird visible; then it sounds higher up, and even in the next tree; all the while the finch is but just overhead, and the moment he moves he is seen. Other birds equally deceive the ear: the yellowhammer does sometimes, and the chattering brook sparrow so will the blackbird when singing – always provided that they are temporarily invisible.

When the crake remains a long time in one place, uttering the call continuously, the illusion disappears, and there is no more difficulty in approximately fixing its position than that of any other bird. One summer a crake chose a spot on the shore of the ditch of the highway hedge, not forty yards from the orchard ha-ha. There was a thick growth of tall grass, clogweed, and other plants just there, and some of the bushes pushed out over the sward. The nest was placed close to the ditch (not in it), and the noise the crakes made was something astonishing. 'Crake, crake, crake!' resounded the moment it was light – and it is light early at that season: 'Crake, crake, crake!' all the morning; the sound now and then, if the bird moved a few yards nearer, echoing back from some of the buildings. There was, or seemed to be, a slight cessation in the middle of the day, but towards evening it recommenced, and continued without cessation till quite dark. This lasted for some weeks: it chanced that the meadow was mown late, so that the birds were undisturbed. Why so apparently timid a bird should choose a spot near a dwelling is not easy to understand.

The crakes, however, when thus localised, deceived no one by their supposed ventriloquial powers; therefore it seems clear that the deception is caused by their rapid changes of position. The mouse in like manner often gives an impression that it must be in one spot when it is really a yard away, the shrill squeak, as it were, left behind it. It is not easy sometimes to fix the position of the deathtick in woodwork. The homefield here is a favourite haunt of the crakes, for like all other birds they have their special places of resort. Another meadow, at some distance on the same farm, is equally favoured by them. This meadow adjoins that second line of bird-travel, following the brook. But as the crakes, though they will take refuge in a hedge, do not travel along it habitually, this circumstance may be accidental. Crakes, notwithstanding they run so swiftly, do not seem to move far when once they have arrived; they appear to restrict themselves to the field they have chosen, or, at the furthest, make an excursion into the next and return again, so that you may always know where to go to hear one.

The mowers cutting these meadows find the eggs – the nest being on the ground – and bring them to the farmstead, both as a curiosity and to be

eaten, some thinking them equal to plovers' eggs. Though you may follow
the sound 'Crake, crake!' in the grass for hours at a time, and sometimes
get so near as to throw your walking stick at a bunch of grass, you will
never see the bird; and nothing, neither stick nor stone, will make it rise.
Yet it is easy to shoot, as I found, in one particular way. The trick is to drive
it into a hedge. Two persons and a spaniel well in hand walk towards the
'Crake, crake!' keeping some distance apart. The bird at first runs straight
away; then, finding himself still pursued, tries to dodge back, but finds the
line extended. He then takes refuge in silence, and endeavours to slip past
unseen and unheard; but the spaniel's power of scent baffles that. At last he
makes for the hedge, when one person immediately goes on the other side,
and the spaniel beats it up. The bird is now surrounded and cannot escape,
and, as the dog comes close upon him, is compelled to rise and fly. As he
rises his flight at first somewhat resembles the partridge's, but it is slower
and heavier, and he can be shot with the greatest ease. But if not fired at,
after he has got well on the wing, the flight becomes much stronger, and it
is evident that he is capable of a long voyage.

Sometimes, by patience and skilfully anticipating his zigzag motions in
the grass, the crake may be driven to the hedge without a dog. He will then,
after a short time, if still hunted, 'quat' in the thickest bunch of grass or
weeds he can find in the ditch, and will stay till all but stepped on, when
he can be knocked down with a walking stick. After the grass is mown, the
crakes leave the meadows and go to the arable fields, where the crops afford

them shelter. This district seems a very favourite resort of these birds.

The mowing-grass while standing does not appear to attract other birds much; but immediately the scythe has passed over they flock to the swathes from the hedges, and come to the hay itself when quite dry. In hay there are many plants whose stems are hollow. Now, as soon as a stalk is dry, if there be any crevice at all, insects will creep in; so that these tiny tubes are frequently full of inhabitants, which probably attract the birds.

Sometimes a bird will perch for a moment on a haymaker's hat as he walks slowly down a lane with hedges each side; the fibres of hay have adhered to it, and the keen eyes above have detected some moving creature on them. Birds that are otherwise timid will remain on the footpath to the very last moment, almost till within reach, if they chance to be dissecting a choice morsel, some exquisite beetle or moth – pecking at it in eager haste and running what to them must seem a terrible risk for the sake of gratifying their taste.

The wood pigeons are fond of acorns, and come for them to the oaks growing in an irregular row along the hedge at the top of the homefield. They are most voracious birds and literally cram their crops with this hard fruit. Squirrels and mice enjoy the nuts in Hazel Corner, and the thrushes and pigeons feed on the 'peggles' which cover the great hawthorn bush there so thickly as to give it a reddish tint. There is a difference even in this fruit: on some bushes the peggles consist mainly of the internal stone, the edible coating being of the thinnest. On others the stone is embedded in a thick mellow covering affording twice as much food. Like other products of the hedge, they are supposed to be improved by frost.

Farther down the highway hedge, by the gateway, a large elder bush, or rather tree, bears a profusion of berries. Blue-black sloes adhere to – they do not hang on – the blackthorn bushes: in places the boughs are loaded with them. Here and there crabs cling to the tough crab tree, whose bark has a dull gloss on it, something like dark polished leather. Bunches of red berries shine on the woodbine; fruit growing in bunches usually depends, but these are often on the upper side of the stalk; and the latter bloom shows by them – flower and fruit at the same time. The berry has a viscous feel.

Larger berries – some red, some green, on the same bunch – cluster on the vines of the bryony. The white bryony, whose leaf is not unlike that of the grape, has a magical reputation, and the cottage folk believe its root to be a powerful ingredient in love potions, and also poisonous. They identify it with the mandrake. If growing in or close to a churchyard its virtues are increased, for, though becoming fainter as they lengthen, the shadows of the old superstitions linger still. Red nightshade berries – not the deadly nightshade, but the bittersweet – hang sullenly among the bushes where this creeping plant has trailed over them. Here and there upon the bank wild gooseberry and currant bushes may be found, planted by birds carrying off ripe fruit from the garden. A wild gooseberry may sometimes be seen growing out of the decayed 'touchwood' on the top of a hollow withy-pollard. Wild apple trees, too, are not uncommon in the hedges.

The beautiful rich colour of the horse chestnut, when quite ripe and fresh from its prickly green shell, can hardly be surpassed; underneath the tree the grass is strewn with the shells, where they have fallen and burst. Close to the trunk the grass is worn away by the restless trampling of horses, who love the shade its foliage gives in summer. The oak apples which appear on the oaks in spring – generally near the trunk – fall off in the summer, and lie shrivelled on the ground not unlike rotten cork, or black as if burned. But the oak galls show thick on some of the trees, light green, and round as a ball; they will remain on the branches after the leaves have fallen, turning brown and hard, and hanging there till the spring comes again.

One of the cottagers in the adjacent hamlet collects these brown balls and strings them upon wire, making flower stands and ornamental baskets for sale. They seem to appear in numbers upon those oak bushes rather than trees which spring when an oak has been cut down but the stump has not been grubbed up. These shoots at first often bear leaves of great size, many times larger than the ordinary oak leaf; some are really immense, measuring occasionally fourteen or fifteen inches in length. As the shoots grow into a bush the leaves diminish in size and become like those of the tree.

In the ditch the tall teasel lifts its prickly head. The large leaves of this plant grow in pairs, one on each side of the stem, and while the plant is young are connected in a curious manner by a green membrane, or continuation of the lower part of the leaf round the stem, so as to form a cup. The stalk rises in the centre of the cup, and of these vessels there are three or four above each other in storeys. When it rains, the drops, instead of falling off as from other leaves, run down these and are collected in the cups, which thus form so many natural rain gauges. If it is a large plant, the cup nearest the ground – the biggest – will hold as much as two or three wine glasses. This water remains there for a considerable time, for several days after a shower, and is fatal to numbers of insects which climb up the stalk or alight on the leaves and fall in. While the grass and the earth of the bank are quite dry, therefore, the teasel often has a supply of water; and when it dries up, the drowned insects remain at the bottom like the dregs of a draught the plant has drained. Round the prickly dome-shaped head, as the summer advances, two circles of violet-hued flowers push out from cells defended by the spines, so that, seen protruding above the hedge, it resembles a tiara – a green circle at the bottom of the dome, and two circles of gems above.

Some of the grasses growing by the hedge are not to be handled carelessly, the edge of the long blade cutting like a lancet: the awn-like seeds of others, if they should chance to get into the mouth, as happens occasionally to the haymakers, work down towards the throat, and the attempt to get rid of them causes a creeping motion the opposite way. This is owing to the awns all slanting in one direction.

On the sultry afternoons of the latter part of the summer the hedge is all but silent. Waiting in the gateway there is no sound for half an hour at a time, no call or merry song in the branches, nothing but the buzz of flies. The birds are quiet, or nearly so: they slip about so noiselessly that it is difficult to observe them, so that many perhaps migrate before it is suspected, and others stay on when thought to be gone. In the grass the grasshoppers make their hiss, and towards evening the yellowhammers utter a few notes; but while the corn is being reaped the meadows are all but still.

TWELVE

The Ash Copse

A GAP IN THE HEDGE by Hazel Corner leads through a fringe of hawthorn bushes into the ash copse. There is a gate at a little distance; but somehow it is always more pleasant to follow the byway of the gap, where two steps, one down into the ditch, or rather on to the heap of sand thrown out from a rabbit bury, and one up on the mound, carry you from the meadow – out of cultivation – into the pathless wood. The green sprays momentarily pushed aside close immediately behind, shutting out the vision, and with it the thought of civilisation. These boughs are the gates of another world. Under trees and leaves – it is so, too, sometimes even in an avenue – where the direct rays of the sun do not penetrate, there is ever a subdued light; it is not shadow, but a light toned with green.

In spring the ground here is hidden by a verdant growth, out of which the anemone lifts its chaste flower. Then the wild hyacinths hang their blue bells so thickly that, glancing between the poles, it is hazy with colour; and in the evening, if the level beams of the red sun can reach them, here and there a streak of imperial purple plays upon the azure. Woodbine coils round the tall straight poles, and wild hops, whose bloom emits a pleasant smell if crushed in the fingers. On the upper and clearer branches of the hawthorn the nightingale sings – more sweetly, I think, in the freshness of the spring morning than at night. Resting quietly on an ash stole, with the scent of flowers and the odour of green buds and leaves, a ray of sunlight yonder lighting up the lichen and the moss on the oak trunk, a gentle air stirring in the branches above, giving glimpses of fleecy clouds sailing in the ether, there comes into the mind a feeling of intense joy in the simple fact of living.

The nightingale shows no timidity while all is still, but sings on the bough in full sight, hardly three yards away, so that you can see the throat swell as the notes are poured forth – now in intricate trills, now a low sweet call, then a liquid 'jug-jug-jug!' To me it sounds richer in the morning – sunlight, flowers, and the rustle of green leaves seem the natural accompaniment; and the distant chorus of other birds affords a contrast and relief – an orchestra filling up the pauses and supporting the solo singer.

Passing deeper into the wood, it is well to be a little careful while stepping across the narrow watercourse that winds between the stoles. Rushes grow thickly by the side, and the slender stream seems to ooze rather than run, trickling slowly down to the brook in the meadow. But the earth is treacherous on its banks – formed of decayed branches, leaves, and vegetable matter, hidden under a thin covering of aquatic grasses. Listen! there is a faint rustling and a slight movement of the grass: it is a snake gliding away to its hole, with yellow-marked head lifted above the ground over which his dull green length is trailing. Stepping well over the moist earth, and reaching the firmer ground, there the thistles grow great and tall, many up to the shoulder; it is a little more open here, the stoles having been cut only two years ago, and they draw the thistles up.

Sometimes the young ash, shooting up after being cut, takes fantastic shapes instead of rising straight. The branch loses its roundness and flattens out to a width of three or four inches, curling round at the top like the conventional scroll ornament. These natural scrolls are occasionally hung up in farmhouses as curiosities. The woodmen jocularly say that the branch grew in the night, and so could not see its way. In some places (where the poles are full-grown) the upper branches rub against each other, causing a weird creaking in a gale. The trees as the wind rises find their voices, and the wood is full of strange tongues. From each green thing touched by its fingers the breeze draws a different note; the bennets on the hillside go 'sish, sish'; the oak in the copse roars and groans; in the firs there is a deep sighing; the aspen rustles. In winter the bare branches sing a shrill 'sir-r-r'.

The elm, with its rough leaf, does not grow in the copse: it is a tree that prefers to stand clear on two sides at least. Oak and beech are here; on their lower branches a few brown leaves will linger all through the winter. Where a huge bough has been sawn from a crooked, ill-grown oak a yellow bloated fungus has spread itself, and under it, if you lift it with a stick, the woodlice are crowded in the rotting stump. The beech boughs seem to glide about, round and smooth, snake-like in their easy curves. The bark of the aspen, and of the large willow poles, looks as if cut with the point of a knife, the cut having widened and healed with a rough scar. On the trunk of the silver birch sometimes the outer bark peels and rolls up of itself. Seen from a distance, the leaves of this tree twinkle as the breeze bends the graceful hanging spray.

The pheasants that wander away from the preserves and covers up under the hills far down in the meadows as the acorns ripen, roost at night here in the copse; and should a storm arise, after every flash of lightning gleaming over the downs the cocks among them crow. So, too, in the daytime, after every distant mutter of thunder the pheasant cocks crow in the preserves, and some declare they can see the flash, even though invisible to human eyes, at noonday.

Clustering cones hang from the firs, fringing the copse on one side – first green, and then a pale buff, and falling at last hard and brown to strew the earth beneath. In the thick foliage of this belt of firs the starlings love to roost. If you should be passing along any road – east, north, west, or south – a mile or two distant, as the sun is sinking and evening approaching, suddenly there will come a rushing sound in the air overhead: it is a flock of starlings flying in their determined manner straight for the distant copse. From every direction these flocks converge upon it: some large, some composed only of a dozen birds, but all with the same intent. If the country chances to be open, the hedges low, and the spectator on a rise so as to see over some distance, he may observe several such flights at the same time. Rooks, in returning to roost, fly in long streams, starlings in numerous separate divisions. This is especially noticeable in summer, when the divisions are composed of fewer birds: in winter the starlings congregate in larger bodies.

It would appear that after the young birds are able to fly they flock together in parties by themselves, the old birds clubbing together also, but all meeting at night. The parties of young birds are easily distinguished by their lighter colour. This may not be an invariable rule (for the birds to range themselves according to age), but it is the case frequently. Viewed from a spot three or four fields away, the copse in the evening seems to be overhung by a long dark cloud-like a bar of mist, while the sky is clear and no dew is yet risen. The resemblance to a cloud is so perfect that anyone – not thinking of such things – may for the time be deceived, and wonder why a cloud should descend and rest over that particular spot. Suddenly, the two ends of the extended black bar contract, and the middle swoops down in the shape of an inverted cone, much resembling a waterspout, and in a few seconds the cloud pours itself into the trees. Another minute and a black streak shoots upwards, spreads like smoke, parts in two, and wheels round back into the firs again.

On approaching it this apparent cloud is found to consist of thousands of starlings, the noise of whose calling to each other is indescribable – the country folk call it a 'charm', meaning a noise made up of innumerable lesser sounds, each interfering with the other. The vastness of these flocks is hardly credible until seen; in winter the bare trees on which they alight become suddenly quite black. Once or twice in the summer starlings may be observed hawking to and fro high in the air, as if imitating the swallows in an awkward manner. Probably some favourite insect is then on the wing, and they resort to this unwonted method to capture it.

Beyond the fir trees the copse runs up into a corner, where hawthorn bushes, briar, and bramble succeed to the ash stoles, and are in turn bordered by some width of furze and brake fern. When this fern is young and fresh the sunshine glistens on its glossy green fronds, but on coming nearer the sheen disappears. On a very hot sultry day towards the end of summer there is occasionally a peculiar snapping sound to be heard in the furze, as if some part of the plant, perhaps the seed, were bursting. The shocks of wheat, too, will crackle in the morning sun. This corner, well sheltered by furze and brake, is one of 'sly Reynard's' favourite haunts. The stems of the furze, when they grow straight, are occasionally cut for walking sticks. Wood pigeons visit the copse frequently – in the spring there are several nests – and towards evening their hollow notes are repeated at intervals. Though without the slightest pretensions to a song, there is something soothing in their call, pleasantly suggestive of woodland glades and deep shady dells.

Just before the shooting season opens there is a remarkable absence of song from hedge and tree: even the chirp of the house sparrow is seldom heard on the roof, where only recently it was loud and continuous. Most of the sparrows have, in fact, left the houses in flocks and resorted to the cornfields after the grain. In this silent season the robin, the wood pigeon, and the greenfinch seem the only birds whose notes are at all common: the pigeons call in the evening as they come to the copse, the greenfinches in a hushed kind of way talk to each other in the hedge, and the robin plaintively utters a few notes on the tree. It is not absolute silence indeed; but the difference is very noticeable. Through the ash poles on one side of the copse distant glimpses may be obtained of gleaming water, where a creek of the shallow lake runs in towards it.

Bordering the furze, a thick hawthorn hedge – a double mound – extends, so wide as to be itself almost another copse. In the 'rowetty' grass on the bank or in the hollow places, under fallen leaves and trailing ivy, the hedgehog hides during the day, so completely concealed that while the sun shines it is extremely difficult to find one without a dog.

A spaniel racing down the mound will pounce on the spot and scratch the hedgehog out in a moment; then, missing the dog, you presently hear a whining kind of bark – half rage, half pain – and know immediately what he is doing. He is trying to unroll the hedgehog, who, as soon as he felt the approach of the enemy, curled himself into a ball, with the sharp spines sticking out everywhere. The spaniel, snapping at the animal, runs these quills deep into his jowl; he draws back, snaps again, shakes his head, and then tries a third time, with bloodspots round his mouth. Every repulse embitters him – his semi-whine expresses intense annoyance, and if left alone there he would stay till covered with blood.

But the older dogs sometimes learn the trick: they then roll the hedgehog over with a paw, touching it gently, so as not to run the spines in, till the depression comes uppermost where the hedgehog has tucked his head

inwards. This is the only vulnerable place, and with one desperate bite the
dog thrusts his teeth in there, seizes the nose, and then has the hedgehog
in his power. The young of the hedgehog are amusing little things, and try
to roll themselves up in precisely the same manner; but they cannot close
the aperture where they tuck the heads in so completely. Though invisible
during the sunshine, hiding so carefully as to be rarely found, when the
dew begins to gather thickly on the grass and the shades deepen they issue
forth, and if you remain quite still show no fear at all. While waiting in
a dry ditch I have often had a hedgehog come rustling slightly along the
bottom till he reached my boot; then he would go up the shore of the ditch
out among the grass, hunting for beetles and the creeping things which he
likes most.

In some places they are numerous; one or two other meadows on the
farm beside the homefield are favourite haunts of theirs, and five or six
may be found out feeding within a short distance. When all is still they
move rapidly through the grass – quite a run; much quicker than they
appear capable of moving. The plough lads, if they find one, carry it to a
pond, knowing that nothing but water will make it unroll voluntarily – no
knocks or kicks; but the moment it touches the water it uncoils. Now and
then a labourer will cook a hedgehog and eat it; some of them will eat a
full-grown rook at any time they chance to shoot it, notwithstanding the
bitter flavour of the bird, only taking out a part of the back. Those who
have had some association with the gypsies seem most addicted to this
kind of food.

In the opposite direction to the ash copse, and about half a mile north
of Wick farmhouse, there rises above the oak and ash trees what looks
like the topmast and yard of a ship lying at anchor or in dock, the hull
hidden by the branches. It is the top of an immensely tall and gaunt fir
tree, whose thin and perhaps dying boughs project almost at right angles.
This landmark, visible over the level meadows for a considerable distance,
stands in that little enclosed meadow as one of the favourite resorts of
birds and wild animals.

From the ash copse the travelling parties come down the highway hedge to the orchard: then, crossing the orchard and road, they enter another thick hedge, which continues in the same general direction; and finally, following it, arrive at this small green mead walled in by trees and mounds so broad as to resemble elongated copses. The mead itself may perhaps be two acres in extent, but it does not appear so much: the part visible on first glancing over the gateway can hardly exceed an acre. The rest is formed of nooks – deep indentations, so to say – not more than six or eight yards wide at the entrance, and running up to a point. Of these there are four or five – recesses in the massive walls of green.

These corners are caused by the mound following the curiously winding course of a brook which flows just on the left side; and on the right side runs a second brook, whose direction is much straighter and current slower. These two meet at the top of the mead, and then, forming a junction, make a deep, swift stream, flowing beside a series of water meadows – broad, level, and open, like a plain – which are irrigated from it. The mounds in the angle where the brooks join enclose a large space planted with osiers, and inside the hedges all round the mead there is a wide, deep ditch, always full of slowly moving water: so that the field is really surrounded by a double moat; and in one corner, in addition, there is a pond hidden by maple thickets from within, and intended for the use of cattle in the adjoining field. The nearest house is several meadows distant, and no footpath passes near, so that the spot is peculiarly quiet. These mounds, hedges, osier bed, and brooks occupy an area nearly or quite equal to the space where cattle can feed.

Upon the fir tree a heron perches frequently in the daytime, because from that great elevation he can command an extensive view, and feels secure against attack. Whenever he visits the water meadows, sailing thither from the shallow lake (one of whose creeks approaches the ash copse), he almost always rests here before descending to the field to take a good look round. The heron is a most suspicious bird: when he alights in the water meadows he stalks about in the very middle of the great field, far out of reach of the gun. If ever he ventures to the brook, it is not till after a careful survey from the fir tree, his tower of observation; and, when in the brook, his long neck

is every now and then extended, that he may gaze above the banks.

By the gateway, reached by crossing a rude bridge for the waggons, wild hops festoon the thickets. Behind the maple bushes in the corner the water of the pond, overhung with willow, is dark – almost black in the depth of shadow. Out of it a narrow and swift current runs into that slow straight brook which bounds the right side of the meadow. Here in the long grass and rushes growing luxuriantly between the underwood lurk the moorhens, building their nests on bunches of rushes against the bank and almost level with the water. Though but barely hatched, and chips of shell clinging to their backs, the tiny fledglings swim at once if alarmed. When a little older they creep about on the miniature terraces formed along the banks by the constant running to and fro of water rats, or stand on a broken branch bent down by its own weight into the water, yet still attached to the stem, puffing up their dark feathers like a black ball.

If all be quiet, the moorhens come out now and then into the meadow;

and then, as they stand upright out of the water, the peculiar way in which their tails, white marked, are turned upwards is visible. The bill is of a fine colour – almost the 'orange-tawny' of the blackbird, set in thick red coral at its base. Under the shallow water at the mouth of the pond the marks of their feet on the mud may be traced: they run swiftly, and depend upon that speed and the skilful tricks they practise in diving – turning back and dodging under water like a hare in the fields – to escape from pursuit, rather than on their wings. Through the thick green flags they creep, and into the hole the water rats have made, or behind and under the natural cavities in the stoles upon the bank. They beat the water with their wings when they rise, showering the spray on either side, for a short distance, and then, ascending on an inclined plane, fly heavily, but with some strength.

Night is their time of journeying, when they come down from the lake or return to it, uttering a weird cry in the darkened atmosphere. By day, as they swim to and fro in the flags and through the duckweed, shaded from the hot sun under willow and aspen, they call to each other, not unpleasantly, a note something like 'croog', with a twirl of the 'r'. In summer they do not move far from the place they have chosen to breed in: in the frosts of winter they work their way up the brooks, or fly at night, but usually come back to the old spot. The dabchick, a slender bird, haunts the pond here too, diving even more quickly than the moorhen.

Nut tree bushes grow along the bank of the brook on this side – the nuts are a smaller sort than usual; and beside the wet ditch within the mound and on the shore, wherever the scythe has not reached, the meadowsweet rears its pale flowers. At evening, if it be sultry, and on some days, especially before a thunderstorm, the whole mead is full of the fragrance of this plant, which lines the inside ditch almost everywhere. So heavy and powerful is its odour that the still motionless air between the thick hedges becomes oppressive, and it is a relief to issue forth into the open fields away from the perfume and the brooding heat. But by day it is pleasant to linger in the shadow and inhale its sweetness – if you are not nervous of snakes, for there is one here and there in the grass gliding away at the jar of the earth under your footstep. Warmth and moisture favour their increase, as on a larger scale in tropic lands; and parts of the mead are often under water when a freshet comes down the brooks so choked with flags that they cannot carry it away quickly.

The osier bed in the angle where the brook joins is on slightly higher ground, for although the withy likes water at its roots it should not stand in it. Springing across the ditch, and entering among the tall slender wands, which, though they look so thick part aside easily, you may find on the mound behind the butt of an oak sawn just above the ground; and there, in the shade of the reeds, and with a cool breeze now and again coming along the course of the stream, it is delicious in the heat of summer to repose and listen to the murmur of the water.

The moorhens come down the current slowly, searching about among the flags; the reed warblers are busy in the hedge; at the mouth of his hole sits a water rat rubbing his face between his paws; across the stream comes his mate, swimming slowly with one end of a long green sedge in her mouth, and the rest towed behind on the surface. They are the beavers of our streams – amusing, intelligent little creatures utterly different in habits from the rat of the drain. Move but a hand, and instantly they fall rather than dive into the water, making a sound like 'thock' as they strike it; and then they run along the bottom, or seem to do so, as swiftly as on dry land. But in a few minutes out they come again, being at the same time extremely timid and as quickly reassured; so that if you remain perfectly still they will approach within a yard.

Where the two brooks meet a hollow willow tree hangs over the brown pool – brown with suspended sand and dead leaves slowly rotating under the surface where the swirl of the meeting currents, one swift and shallow, the other deeper and stronger, has scooped out a basin. A waving line upon the surface marks where the two streams shoulder each other and strive for mastery, and its curve, yielding now to this side, now to that, responds to their varying volume and weight. While the undercurrents sweep ever slowly round, whirling leaf and dead, black, soddened twigs over the hollow, the upper streams are forced together unwillingly by the narrowing shores, and throw themselves with a bubbling rush onwards.

Through the brown water, from under the stooping willow whose age bows it feebly, there shine now and again silvery streaks deep down as the roach play to and fro. There, too, come the perch; they are waiting for the insects falling off the willows and the bushes, and for the food brought down by the streams.

'Hush!' it is the rustle of the reeds, their heads are swaying – a reddish brown now, later on in the year a delicate feathery white. Seen from beneath, their slender tips, as they gracefully sweep to and fro, seem to trace designs upon the blue dome of the sky. A whispering in the reeds and tall grasses; a faint murmuring of the waters: yonder, across the broad water meadow, a yellow haze hiding the elms.

In the nooks and corners on the left side of the mead the hemlock rears its sickly-looking stem; the mound is broad and high, and thickly covered with grasses, for the most part dead and dry. These form a warm cover for the fox: there is usually one hiding somewhere here, the mead being so quiet. Where the ground is often flooded watercress has spread out into the grass, growing so profusely that now the water is low it might be mown by the scythe. And everywhere in their season the beautiful forget-me-nots nestle on the shores among the flags, where the water, running slower at the edge, lingers to kiss their feet.

Once, some five-and-twenty years ago, a sportsman startled a great bird out of the spot where the streams join, and shot it, thinking it was a heron. But seeing that it was no common heron, he had it examined, and it was found to be a bittern, and as such was carefully preserved. It was the last visit of bitterns to the place; even then they were so rare as not to be recognised: now the progress of agriculture has entirely banished them.

THIRTEEN

The Warren

UNDER THE TRUNKS of the great trees the hedges are usually thinner, and need repairing frequently; and so it happens that at the top of the homefield, besides the gap leading into the ash copse, there is another some distance away beneath a mighty oak. By climbing up the mound, and pushing through the brake fern which grows thickly between the bushes, entrance is speedily gained to the wide rolling stretch of open pasture called The Warren. The contrast with the small enclosed meadow just left is very striking. A fresh breeze comes up from the lake, which, though not seen in this particular spot, borders the plain-like field in one part.

The ground is not level; it undulates, now sinking into wide hollows, now rising in rounded ridges, and the turf (not mown but grazed) is elastic under the foot, almost like that of the downs in the distance. This rolling surface increases the sense of largeness – of width – because it is seldom possible to see the whole of the field at once. In the hollows the ridges conceal its real extent: on the ridges a corresponding rise yonder suggests another valley. The two rows of tall elms – some hundreds of yards apart – the scattered hawthorn bushes and solitary trees, groups of cattle in the shade, and sheep grazing by the far-away hedge, give the aspect of a wilder park, the more pleasant because of its wildness.

Near about the centre, where the land is most level, an unexpected slope goes down into a cuplike depression. This green crater may perhaps have been formed by digging for sand – so long ago that the turf has since grown over smoothly. Standing at the bottom the sides conceal all but the sky overhead. Some few dead leaves of last year, not yet decayed, though bleached and brittle, lie here at rest from the winds that swept them over the plain. Silky balls of thistledown come irresolutely rolling over the hedge, now this way, now that: some rise and float across, some follow the surface and cling a while to the bennets in the hollow. Pale blue harebells, drooping from their slender stems here and there, meditate with bowed heads, as if full of tender recollections.

Now, on hands and knees (the turf is dry and soft), creep up one side of the bowl-like hollow, where the thistles make a parapet on the edge, and from behind it look out upon the ground all broken up into low humps, some covered with nettles, others heaps of sand: it is the site of an immense rabbit burrow, the relic of an old warren which once occupied half the field. The nettle-covered heaps mark old excavations; where the sand shows, there the miners have been recently at work. At the sound of approaching footsteps those inhabitants that had been abroad hastily rushed into their caves, but now (after waiting a while, and forgetting that the adjacent hollow might hide the enemy) a dozen or more have come forth within easy gunshot. Though a few like this are always looking in and out all through the day, it is not till the approach of evening that they come out in any number.

This is a favourite spot from whence to get a shot at them, but the aim must be deadly, or the rabbit will escape though never so severely wounded. The holes are so numerous that he has never more than a yard to scramble, and as he goes down into the earth his own weight carries him on. If he can but live ten seconds after the lead strikes him, he will generally escape you. Watching patiently (without firing), after the twilight has deepened into night, presently you are aware of a longer, larger creature than a rabbit stealing out, seeming to travel close to the earth: it is a badger. There are almost always a couple somewhere about the warren. Their residence is easily discovered because of the huge heap of sand thrown out from the rabbit hole they have chosen; and it is this ease of discovery that has caused the diminution of their numbers by shot or spade.

The ground sounds hollow underneath the foot – perhaps half an acre is literally bored away under the surface; and you have to thread your way in and out a labyrinth of holes, the earth about some of them perceptibly yielding to your weight. There must be waggonloads of the sand that has been thrown out. Beyond this central populous quarter suburbs of burrows extend in several directions, and there are detached settlements fifty and a hundred yards away. In ferreting this place the greatest care has to be taken that the ferret is lined with a long string, or so fed that he will not lie in; otherwise, if he is not picked up the moment he appears at the mouth of the hole, he will become so excited at the number of rabbits, and so thirsty for blood, that he will refuse to come back.

To dig for him is hopeless in that catacomb of tunnels; there is nothing for it but to send a man day after day to watch, and if possible to seize him while passing along the upper ground from one bury to another. In time thirst will drive him to wander; there is no water near this dry, sandy, and rather elevated spot, and blood causes great thirst. Then he will roam across the open, and by and by reach the hedges, where in the ditch some water is sure to be found in winter, when ferreting is carried on. So that if a ferret has been lost some time, it is better to look for him round the adjacent hedges than in the warren.

Long after leaving the bury it is as well to look to your footsteps, because of solitary rabbit holes hidden by the grass growing up round and even over them. If the foots sinks unexpectedly into one of these, a sprained ankle or even a broken bone may result. Most holes have sand round the mouth, and may therefore be seen even in the dusk; but there are others also used which have no sand at the mouth, the grass growing at the very edge. Those that have sand have been excavated from without, from above; those that have not have been opened from below. The rabbit has pushed his way up from an old bury, so that the sand he dug fell down behind him into the larger hole.

The same thing may be seen in banks, though then the holes worked from within are not so much concealed by grass. These holes are always very much smaller than the others, some so small that one might doubt how a rabbit could force his body through them. The reason why the other tunnels appear so much larger is because the rabbit has no means of shoring up his excavations with planks and timbers, and no 'cage' with which to haul up the sand he has moved; so that he must make the mouth wider than is required for the passage of his body, in order to get the stuff out behind him. He can really creep through a much smaller aperture. At night especially, when walking near a bury situated in the open field, beware of putting your foot into one of these holes, which will cause an awkward fall, if nothing worse. Some of the older holes, now almost deserted, are, too, so hidden by nettles and coarse grass as to be equally dangerous.

The hereditary attachment of wild animals for certain places is very noticeable at the warren. Though annually ferreted, shot at six months out of the twelve, and trapped – though weasels and foxes prey on the inhabitants – still they cling to the spot. They may be decimated by the end of January, but by September the burrows are as full as ever. Weasels and stoats of course come frequently, bent on murder, but often meet their own doom through over-greediness; for some one generally comes along with a gun once during the day, and if there be any commotion among the rabbits, waits till the weasel or stoat appears at the mouth of a hole, and sends a charge of shot at him. These animals get caught, too, in the gins, and altogether would do better to stay in the hedgerows.

The grass of this great pasture has a different appearance to that in the meadows which are mown for hay. It is closer and less uniformly green, because of the innumerable dead fibres. There are places which look almost white from the bennets which the cattle leave standing to die after the seeds have fallen, and shrink as their sap dries up. Somewhat earlier in the summer, bright yellow strips and patches, like squares of praying-carpet thrown down upon the sward, dotted the slopes: it was the bird's-foot lotus growing so thickly as to overpower the grass. Mushrooms nestle here and there: those that grow in the open, far from hedge and tree, are small, and the gills of a more delicate salmon colour. Under the elms yonder a much larger variety may be found, which, though edible, are coarser.

Where a part of the lake comes up to the field is a long-disused quarry, whose precipices face the water like a cliff. Thin grasses have grown over the excavations below: the thistles and nettles have covered the heaps of rubbish thrown aside. The steep, inaccessible walls of hardened sand are green with minute vegetation. Along the edge above runs a shallow red-brown band – it is the soil which nourishes the roots of the grasses of the field: beneath it come small detached stones in sand; these fall out, loosened by the weather, and roll down the precipice. Then, still deeper, the sand hardens almost into stone, and finally comes the stone itself; but before the workmen could get out more than a thin layer they reached the level of the water in the lake, which came in on them, slowly forming pools.

These are now bordered by aquatic grasses, and from their depths every now and then the newts come up to the surface. In the sand precipices are small round holes worked out by the martins – there must be scores of them. Where narrow terraces afford access to four-footed creatures, the rabbits, too, have dug out larger caves; some of them rise upwards, and open on the field above, several yards from the edge of the cliff. The sheep sometimes climb up these ledges; they are much more active than they appear to be, and give the impression that in their native state they must have rivalled the goats. The lambs play about in dangerous-looking places without injury: the only risk seems to be of their coming unexpectedly on the cliff from above; if they begin from below they are safe. A wood pigeon may frequently be found in the quarry – sometimes in the pits, sometimes on the ledges high up – and the goldfinches visit it for the abundant thistledown.

Between the excavated hollow and the lake there is but a narrow bank of stone and sand overgrown with sward; and, reclining there, the eye travels over the broad expanse of water, almost level with it, as one might look along a gun barrel. Yonder the roan cattle are in the water up to their knees; the light air ripples the surface, and the sunshine playing on the wavelets glistens so brilliantly that the eye can scarcely bear it; and the cattle ponder dreamily, standing in a flood of liquid gold.

A path running from Wick across the fields to the distant downs leads to the forest. It would be quite possible to pass by the edge without knowing that it was so near, for a few scattered trees on the hillside would hardly attract attention. Nothing marks where the trees cease: thin, wide apart, and irregularly placed, because planted by nature, they look but a group on the down. There is indeed a boundary, but it is at a distance and concealed: it is the trout stream in the hollow far below, winding along the narrow valley, and hidden by osier beds and willow pollards.

Ascending the slope of the down towards the trees, the brown-tinted grass feels slippery under foot: this wiry grass always does feel so as autumn approaches. A succession of detached hawthorn bushes – like a hedge with great gaps – grow in a line up the rising ground – the dying vines of the bryony trail over them – one is showing its pale greenish-white flowers, while the rest bear heavy bunches of berries. A last convolvulus, too, has a single pink-streaked bell, though the bough to which it holds is already partly bare of leaves. The touch of autumn is capricious, and passes over many trees to fix on one which stands out glowing; with colour, while on the rest a dull green lingers. Near the summit a few bunches of the brake fern rise out of the grass; then the foremost trees are reached, beeches as yet but faintly tinted here and there. Their smooth, irregularly round trunks are of no great height – both fern and trees at the edge seem stunted, perhaps because they have to bear the brunt and break the force of the western gales sweeping over the hills.

For the first two hundred yards the travelling is easy because of this very scantiness of the fern and underwood; but then there seems to rise up a thick wall of vegetation. To push a way through the ever-thickening bracken becomes more and more laborious; there is scarce a choice but to follow a winding narrow path, green with grass and moss, and strewn with leaves, in and out and round the impenetrable thickets. Whither it leads – if, indeed, anywhere – there is no sign. The precise sense of direction is quickly lost, and then glancing round and finding nothing but fern and bush and tree on every hand, it dawns upon the mind that this is really a forest – not a wood, where a few minutes either way will give you a glimpse of the outer light through the ash poles.

Other narrow paths – if they can be called paths which show no trace of human usage – branch off from the original one till by-and-by it becomes impossible to recognise one from the other. The first has been lost indeed long ago, without its having been observed: for the bracken is now as high as the shoulders, and the eye cannot penetrate many yards on either side. Under a huge oak at last there is an open space, circular, and corresponding with the outer circumference of its branches: carpeted with dark green grass and darker moss, thickly strewn with brown leaves and acorns that have dropped from their cups. A wall of fern encloses it: the path loses itself in the grass because it is itself green.

Several such paths debouch here – which is the right one to follow? It is pure chance. On again, with more tall bracken, thorn thickets, and maple bushes, and noting now the strange absence of living things. Not a bird rises startled from the boughs, not a rabbit crosses the way; for in the forest, as in the fields, there are places haunted and places deserted, save by occasional passing visitors. Suddenly the bracken ceases, and the paths disappear under a thick grove of beeches, whose dead leaves and beechmast seem to have smothered vegetation.

Insensibly the low ground rises again, the brake and bushes and underwood reappear, but the trees grow thinner and farther apart; they are mainly oaks, which like to stand separate in their grandeur. There is one dead oak all alone in the midst of the underwood, with a wide space around it. A vast grey trunk, split and riven and hollow, with a single pointed branch rising high above it, dead too, and grey: not a living twig, not so much as a brown leaf, gives evidence of lingering life. The oak is dead; but even in his death he rules, and the open space around him shows how he once overshadowed and prevented the growth of meaner trees. More oaks, then a broad belt of beeches, and out suddenly into an opening.

It is but a stone's throw across – a level mead walled in with tall trees, whose leaves in myriads lie on the brown tinted grass. One great thicket only grows in the midst of it. The nights are chilly here, as elsewhere; but in the day, the winds being kept off by the trees and underwood, it becomes quite summer-like, and the leaves turn to their most brilliant hues. The stems of the bracken are yellow; the fronds vary from pale green and gold, commingled, to a reddish bronze. The hawthorn leaves are slight yellow, some touched with red, others almost black. Maple bushes glow with gold. Here the beeches show great spots of orange, yonder the same tree, from the highest branch to the lowest, has become a rich brown. Brown, too, and buff are the oaks; but the tints so shade into each other that it is hard to separate and name them.

It is not long before sounds and movements indicate that the forest around is alive with life. Often it happens that more may be observed while stationary in one spot than while traversing a mile or two; for many animals crouch or remain perfectly still, and consequently invisible, when they hear a footstep. There is a slight tapping sound – it seems quite near, but it is really some way off; and presently a woodpecker crosses the open, flying with a wave-like motion, now dipping and now rising. Soon afterwards a second passes: there are numbers of them scattered about the forest. A clattering noise comes from the trees on the left – it is a wood pigeon changing his perch; he has settled again, for now his hollow note is heard, and he always calls while perching. A loud screeching and chattering deeper in the forest tells that the restless jays are there. A mistle thrush comes and perches on a branch right overhead, uttering his harsh note, something like turning a small rattle. But he stays a moment only: he is one of the most suspicious of birds, and has instantly observed that there is someone near. A magpie crosses the mead and disappears.

Something moving yonder in the grass catches the eye; it is a reddish bushy tail, apparently without a body, yet held nearly upright, and moving hither and thither in a quick, nervous way. Suddenly down it goes, and the squirrel raises himself on his haunches to listen to some suspicious sound, holding his forefeet something like a kangaroo. Then he recommences searching and the tail rises, alone visible above the tall grass. Now he bounds, and as his body passes through the air the tail extends behind and droops so that he seems to form an arch. After working along ten or fifteen yards in one direction, he stops, turns sharp round, and comes all the way back again. Some distance farther, under the trees, two more are frisking about, and a rabbit has come to nibble the grass in the open.

Looking across to the other side, where the fern recommences, surely there was a movement as if a branch was shaken; and a branch that, on second thought, is in such a position that it cannot be connected with any tree. Again, and then the head and neck of a stag are lifted above the fern.

He is attacking a tree – rubbing his antlers against a low branch, much as if he were fighting it. He is not a hundred yards off; it would be easy to get nearer, surely, by stalking him carefully, gliding from tree trunk to tree trunk under the beeches.

At the first step the squirrel darts to the nearest beech; and although it seems to have no boughs or projections low down, he is up it in a moment, going round the trunk in a spiral. A startling clatter resounds overhead: it is a wood pigeon that had come quietly and settled on a tree close by, without being noticed, and now rises in great alarm. But it is a sound to which the deer are so accustomed that they take no notice. There is a little underwood here beneath the beeches, but the beechmast lies thick, and there are dead branches, which if stepped on will crack loudly.

A weasel rushes past almost underfoot; he has been following his prey so intently as not to have observed where he was going. He utters a strange, startled 'yap', or something between that and the noise usually made by the lips to encourage a horse, and makes all speed into the fern. These are happy hunting grounds for weasels.

During spring and summer – so long as the grass, clover, and corn crops are standing, and are the cover in which partridges and other birds have their nests – the weasels and stoats haunt the fields, being safe from observation (while in the crops) and certain of finding a dinner. Then, if you watch by a gap in the hedge, or look through a gateway into the cornfield, you may be almost certain of seeing one at least; in a morning's walk in summer I have often seen two or three weasels in this way. The young rabbits and leverets are of course their prey also. But after the corn is cut you may wait and watch a whole day in the fields and not see a weasel. They have gone to the thick mounds, the covers, woods, and forests, and therein will hunt the winter through.

The stag is still feeding peacefully; he is now scarce fifty yards away, when he catches sight and is off. His body as he bounds seems to keep just above the level of the fern. It is natural to follow him, though of course in vain; the mead is left behind, and once more there is a wall of fern on either side of the path. After a while a broad green drive opens, and is much more

easy to walk along. But where does it go? For presently it divides into two, and then the fork pursued again branches. Hush! What is that clattering? It sounds in several directions, but nothing is visible.

Then a sharp turn of the drive opens on a long, narrow, grassy valley, which is crowded with deer. Parties of thirty or forty are grazing; and yonder, farther away by themselves, there must be nearly a hundred fawns. Standing behind a tree, it is a pleasant sight to watch them; but after a while comes back the thought, beforehand dismissed contemptuously: the afternoon is advancing and is it possible to be lost? The truth is we are lost for the time.

It is impossible to retrace one's footsteps, the paths and drives are so intricate, and cross and branch so frequently. There are no landmarks. Perhaps from the rising ground across the valley a view may be obtained. On emerging into the open, the whole herd of deer and fawns move slowly into the forest and disappear. From the hill there is nothing visible but trees. If a tree be climbed to get a lookout, there is still nothing but trees. Following a green drive as a forlorn hope, there comes again the rattling as of clubs and spears, and strange grunting sounds. It is the bucks fighting; and they are not altogether safe to approach. But time is going on; unless we can soon discover the way, we may have to remain till the tawny wood-owls flit round the trees.

There comes the tinkle-tinkle of a bell: a search shows two or three cows, one of which, after the fashion of the old time, carries a bell. She comes and butts one playfully, and insists on her poll being rubbed. Then there is more grunting, but of a different kind – this time easily recognised: it is a herd of swine searching for the beechmast and acorns. With them, fortunately, comes the swineherd – a lad, who shows a drive which leads to the nearest edge of the forest.

Half an hour after leaving the swineherd, a rabbit is found sitting on his haunches, motionless, with the head drooping on one side. He takes no notice – he is dying. Just beneath one ear is a slight trace of blood – it is the work of a weasel, who fled on hearing approaching footsteps. Soon a film must form over the beautiful eye of the hunted creature: let us in mercy strike him a sharp blow in the head with the heavy end of the walking stick, and so spare him the prolonged sense of death. A hundred yards farther is a gate, and beyond that an arable field. On coming near the gate a hawk glides swiftly downwards over the hedge that there joins the forest. A cloud of sparrows instantly rise from the stubble, and fly chirping in terror to the hedge for shelter; but one is too late, the hawk has him in his talons. Yonder is a row of wheat ricks, the fresh straw with which they have just been covered contrasting with the brown thatch of the farmhouse in the hollow. There a refreshing glass of ale is forthcoming, and the way is pointed out.

FOURTEEN

The Rookery

THE CITY BUILT BY THE ROOKS in the elms of the great pasture field (the Warren, near Wick farmhouse) is divided into two main parts; the trees standing in two rows, separated by several hundred yards of sward. But the inhabitants appear to be all more or less related, for they travel amicably in the same flock and pay the usual visit to the trees at the same hour. Some scattered elms form a line of communication between the chief quarters, and each has one or more nests in it. Besides these, the oaks in the hedgerows surrounding the field support a few nests, grouped three or four in close neighbourhood. In some trees near the distant ash copse there are more nests, whose owners probably sprang from the same stock, but were exiled, or migrated, and do not hold much communion with the capital.

In early days men seem to have frequently dug their entrenchments or planted their stockades on the summit of hills. To the rooks their trees are their hills, giving security from enemies. The wooden houses in the two main streets are evidently of greater antiquity than those erected in the outlying settlements. The latter are not large or thick: they are clearly the work of one, or at most two, seasons only; for it is noticeable that when rooks build at a distance from the centre of population they are some time before they finally decide on a site, abandoning one place after another. But the nests forming the principal streets are piled up to a considerable height – fresh twigs being added every year – and are also thick and bulky. The weight of the whole must be a heavy burden to the trees.

Much skill is shown in the selection of the branches upon which the foundations are laid. In the first place, the branch must fork sufficiently to hold the bottom twigs firmly and to give some side-support. Then it must be a branch more or less vertical, or it would swing with the wind too much up and down as well as to and fro. Thirdly, there should be a clear or nearly clear space above the nest to give easy access, and to afford room for it to increase in size annually. For this reason, perhaps, nests are generally placed near the top or outer sides of the tree, where the boughs are smaller, and every upward extension reaches a clearer place. Fourthly, the bough ought not to be too stiff and firm; it should yield a little, and sway easily, though only in a small degree, to the breeze. If too stiff, in strong gales the nest runs the risk of being blown clean out of the tree. Fifthly, no other branch must rub against the one bearing the principal weight of the nest, for that would loosen the twigs in time, and dislocate the entire structure. Finally, rooks like an adjacent bough on which the bird not actually engaged in incubation can perch and 'caw' to his mate, and which is also useful to alight on when bringing food for the young.

It may be that the difficulty of finding trees which afford all these necessary conditions is one reason why rooks who settle at a distance from their city seem long before they can please themselves. The ingenuity exercised in the selection of the bough and in the placing of the twigs is certainly very remarkable. When the wind blows furiously you may see the nest moving gently, riding on the swaying boughs, while one of the birds perches on a branch close by, and goes up and down like a boat on the waves. Except by the concussion of branches beating hard against the nest, it is rarely broken; up to a certain point it would seem as if the older nests are the firmest, perhaps because of their weight. Sometimes one which has been blown down in the winter – when the absence of protecting leaves gives the wind more power on them – retains its general form even after striking against branches in its descent and after collision with the earth.

Elms are their favourite trees for building in. Oak and ash are also used, but where there are sufficient elms they seem generally preferred. These trees, as a rule, grow higher than any others ordinarily found in the fields, and are more frequently seen in groups, rows, or avenues, thus giving the rooks facilities for placing a number of nests in close neighbourhood. The height of the elm affords greater safety, and the branches are perhaps better suited for their purpose.

After building in an elm for many years – perhaps ever since the owner can remember – rooks will suddenly desert it. There are the old nests still; but no effort is made to repair them, and no new ones are made. The winds and storms presently loosen the framework, about which no care is now taken, and portions are blown down. Then by-and-by the discovery is made that the tree is rapidly dying. The leaves do not appear, or if they do they wither and turn yellow before Midsummer: gradually the branches decay and fall of their own weight or before the wind.

No doubt if anyone had carefully examined the tree he would have observed signs of decay long before the rooks abandoned it; but those who pass the same trees day after day for years do not observe minute changes, or, if they do, as nature is slow in her movements, get so accustomed to the sight of the fungi about the base, and the opening in the bark where the decomposing touchwood shows, as to think that it will always be so. At last the rooks desert it, and then the truth is apparent.

Their nests, being heavy, are not safe on branches up which the strengthening sap no longer rises; and in addition to the nest there is the weight of the sitting bird, and often that of the other who perches temporarily on the edge. As the branches die they become stiff, and will not bend to the gale; this immobility is also dangerous to the nest. So long as the bough yields and sways gently – not much, but still a little – the strong winds do no injury. When the bough becomes rigid, the broad side or wall of the nest offers an unyielding surface, which is accordingly blown away.

The nests which contain young are easily distinguished, despite the height, by the almost continuous cry for food. The labour of feeding the voracious creatures must be immense, and necessity may partly account for the greater boldness of the old birds at that season. By counting the nests from which the cry proceeds the condition of the rookery is ascertained, and the amount of sport it will afford reckoned with some certainty. By noting the nests from which the cry arose last, it is known which trees to avoid in the rook shooting; for the young do not all come to maturity at the same time, and there are generally a dozen or so which it is best to have a week or a fortnight later than the rest.

When the young birds begin to quit the nests, and are observed perching on the tree or fluttering from branch to branch, they must not be left much longer before shooting, or they will wander and be lost. A very few days will then make all the difference; and so it has often happened that men expecting to make a great bag have been quite disappointed, notwithstanding the evident number of nests; the shooting has been held a day or so too late. The young birds get the use of their wings very quickly, and their instinct rather seems to be to wander than to remain in the immediate vicinity of their birthplace.

Some think that the old birds endeavour to entice them away as much as possible, knowing what is coming. It may be doubted if that is the case with respect to the very young birds; but when the young ones are capable of something like extended flight, and can cross a field without much difficulty, I think the parents do attempt to lead them away. When the shooting is in progress, if you will go a little distance from the rookery, out of the excitement of the sport, you may sometimes see two old rooks, one on each side of a young one, cawing to it with all their might. The young bird is, perhaps, on the ground, or on a low hedge, and the old birds are evidently endeavouring to get it to move. Yet they have not learned the only way in which that can be done – i.e. by starting themselves and flying a short distance, and waiting, when the young bird will almost invariably follow.

If you approach the trio the two old birds at once take flight, seeing your gun, and the young bird in a few seconds goes after them. Had they the sense to repeat this operation, they might often draw the young one away from danger; as for their cawing, it does not seem to be quite understood by their offspring, who have hardly yet learned their own language.

To appreciate this effort on the part of the old birds, it must be recollected that immediately after the first shot the great mass of the old rooks fly off in alarm. They go to some distance and then wheel round and come back at an immense height, and there, collected in loose order, circle round and round, cawing as they sail. For an old rook to remain in or near the rookery when once the firing has commenced is the exception, and must be a wonderful effort of moral courage, for of all birds rooks seem most afraid of a gun; and naturally so, having undergone, when themselves young, a baptism of fire. Those that escape slaughter are for the most part early birds that come to maturity before the majority, and so leave the trees before the date fixed for shooting arrives, or acquire a power of flight sufficient to follow their parents on the first alarm to a safe distance. They have, therefore, a good opportunity of witnessing the destruction of their cousins, and do not forget the lesson.

Although the young birds upon getting out of the nest under ordinary conditions seem to like to wander, yet if they are driven out or startled by the shot they do not then at once endeavour to make for the open country or to spread abroad, but appear rather to cling to the place, as if the old nests could shelter them. After a while they begin to understand the danger of this proceeding, and half an hour's rapid firing causes the birds to spread about and get into the trees in the hedges at some distance. There of course they are pursued, or killed the next day, three-quarters of a mile or more away from home. It is rare for old rooks to get shot, for the reason above stated: they rise into the air out of reach. Those that are killed are generally such as have lingered in the hope to save a young bird, and are mistaken and shot as young themselves.

Young birds may be easily distinguished by their slow, uncertain flight and general appearance of not knowing exactly where to go or what to do. They are specially easy to pick out if you see them about to perch on a tree. They go at the tree anyhow, crash in among the branches, and rather fall on a perch than choose it. The old bird always enters a tree carefully, as if he did not like to ruffle his feathers, and knew precisely what sort of bough he preferred to settle on. Close to the rookery there is no need to wait to pick out the young birds, because they are all sure to be young birds there; but, as observed, old birds will linger with young ones at a little distance, and may then be mistaken – as also on the following day, when sportsmen go round to pick up the outsiders, and frequently come on old and young together. The old bird will not sit and let you aim at him perching; if you shoot him, it must be on the wing. The young bird will sit and let you pick him off with a crossbow, and even if a cartridge singes his wing he will sometimes only hop a yard or two along the boughs.

Though hard hit and shattered with shot, they will cling to the branches convulsively, seeming to hang by the crook of the claw or by muscular contraction even when perfectly dead, till lifted up by a shot fired directly underneath, or till the bough itself is skilfully cut off by a cartridge and both come down together. The young feathers being soft, and the quills not so hard as in older birds, scarcely a rook shooting ever goes by without some one claiming to have made a tremendous long shot, which is quite possible, as it does not require many pellets or much force behind them.

On dropping a rook, probably at some distance from the rookery, where the men are whose duty it is to collect the slain, beware of carrying the bird; let him lie, or at most throw him upon a bramble bush in a conspicuous spot till a boy comes round. Rooks are perfectly infested with vermin, which in a few minutes will pass up their legs on to your hand, and cause an unpleasant irritation, though it is only temporary – for the insects cannot exist long away from the bird.

The young birds are occasionally stolen from the nests, notwithstanding the difficulty of access. Young labourers will climb the trees, though so large that they can scarcely grasp the trunk, and with few branches, and those small for some height; for elms are often stripped up the trunk to make the timber grow straight and free from the great branches called 'limbs'. Even when the marauder is in the tree he has some difficulty in getting at the nests, which are placed where the boughs diminish in size – climbing irons used to be sometimes employed for the purpose. As elm trees are so conspicuous, these thieving practices cannot well be carried on while it is light. So the rook poachers go up the trees in the dead of night; and as the old rooks would make a tremendous noise and so attract attention, they carry a lantern with them, the light from which silences the birds. So long as they can see a light they will not caw.

The time selected to rob a rookery is generally just before the date fixed for the shooting, because the young birds are of little use for cooking till ready to fly. The trick, it is believed, has often been played, for the mere pleasure of spiting the owner, the very night previous to the rook shooting party being chosen. These robberies of young rooks are much less frequent than they used to be. One reason why those who possess any property in the country do not like to see a labouring man with a gun is because he will shoot an old rook (and often eat it), if he gets the opportunity, without reference to times or seasons, whether they are building or not.

The young rooks that escape being shot seem to be fed, or partly fed, by the old birds for some time after they can fly well and follow their parents. It is easy to know when there are young rooks in a flock feeding in a field. At the first glance the rooks look scattered about, without any order, each independent of the other. But in a few minutes it will be noticed that here and there are groups of three, which keep close together. These are formed of the parents and the young bird – apparently as big and as black as themselves – which they feed now and then. The young bird, by attending to their motions, learns where to find the best food. As late as July trios like this may sometimes be seen.

Besides the young birds that have the good fortune to pass unscathed through the dangers of rook shooting day, and escape being knocked over afterwards, some few get off on account of having been born earlier than the majority, thus possessing a stronger power of flight. Some nests are known to be more forward than the others; but although the young birds may be on the point of departing, they are not killed, because the noise of the firing would disturb the whole settlement. So that it becomes the rook's interest to incubate a little in advance of the rest.

After a few months they are put into another terrible fright – on the first of September. Guns are going off in all directions, no matter where they turn, so that they find it impossible to feel at ease, and instead of feeding wheel about in the air, or settle on the trees.

The glossy plumage of the rook will sometimes, when seen at a certain angle, reflect the sun's rays in such a manner that instead of looking black the bird appears clothed in shining light: it is as if the feathers were polished like a mirror. In feeding they work in a grave, steady way – a contrast to the restless starlings who so often accompany them. They do not put a sentinel in a tree to give warning of the approach of an enemy. The whole flock is generally on the ground together, and if half a dozen perch awhile on the trees they soon descend. So far are they from setting a watch, that if you pass up outside the hedge to the leeward, on any side except where the wind would carry the noise of footsteps to them, it is easy to get close – sometimes, if they are feeding near the hedge, within three or four yards. Of course, if a rook happens to be in a tree it will not be possible to do so; but they do not set a sentinel for this purpose.

Rooks, in a general way, seem more at their ease in the meadows than in the arable fields. In the latter they are constantly fired at, if only with blank charges, to alarm them from the seed, besides being shouted at and frightened with clappers. The birdkeeper's efforts are, however, of very little avail. If he puts the flock up on one side of the field, they lazily sail to a distant corner, and when he gets there go back again. They are fully aware that he cannot injure them if they keep a certain distance; but this perpetual driving to and fro makes them suspicious. In the meadows it is rare for them to be shot at, and they are consequently much less timid.

At the same time they can perfectly well distinguish a gun from a walking stick. If you enter a meadow with a gun under your arm, and find a flock feeding, they immediately cease searching for food and keep a strict watch on your movements; and if you approach they are off directly. If you carry a walking stick only, you may pass within thirty yards sometimes, and they take little notice, provided you use the stick in the proper way. But now lift it, and point it at the nearest rook, and in an instant he is up with a 'caw' of alarm – though he knows it is not a gun – and flies just above the surface of the ground till he considers himself safe from danger. Often the whole flock will move before that gesture. It is noticeable that no wild creatures, birds or animals, like anything pointed at them; you may swing your stick freely, but point it, and off goes the finch that showed no previous alarm. So, too, dogs do not seem easy if a stick is pointed at them.

Rooks are easily approached in the autumn, when gorging the acorns. They may often be seen flying carrying an acorn in the bill. Sometimes a flock will set to work and tear up the grass by the roots over a wide space

– perhaps nearly half an acre – in search of a favourite beetle. The grass is pulled up in little wisps, just about as much as they can hold in their beaks at a time. In spring they make tracks through the mowing-grass – not in all the meadows, but only in one here and there, where they find the food they prefer. These tracks are very numerous, and do the grass some damage. Besides following the furrows made by the plough, and destroying grubs, beetles, wireworm, and other pests in incalculable numbers, they seem to find a quantity of insect food in unripe corn; for they often frequent wheat fields only just turning yellow, and where the grain is not yet developed. Except perhaps where they are very numerous, they do much more good than harm.

Rooks may now and then be seen in the autumn, on the hayricks; they pull the thatch out, and will do in this way an injury to the roof. Therefore old black bottles are often placed on the thatch in order to scare them. It is said that they pull out the straw for the stray grains left in the ear by the thrashing machine. This seems doubtful. It appears more probable that some insect found on the straw attracts them.

If you are walking past a feeding flock, the nearest rook to you will often exhibit a ridiculous indecision as to whether he shall fly or not. He stretches his neck and leans forward as if about to spring, stops, utters a questioning 'Cawk?' then watches you a moment and gives a hop, just opens his wings, shuts them, and descends within a couple of feet. 'Cawk!' again. Finally, if you turn from your course and make a step towards him, he rises, flaps his wings three or four times, extends them, and glides a dozen yards to alight once more.

Sometimes a flock will rise in the air, and silently wheel round and round after each other, gradually ascending and drifting slowly with the current till they reach a great height. When they soar like this it is said to foretell fine weather. At another time a flock will go up and wheel about in the strangest irregular manner. Every now and then one will extend his wings, holding them rigid, and dive downwards in his headlong descent wavering to and fro like a sheet of paper falling edge first. He falls at a great pace, and looks as if he must be dashed to pieces against a tree or the earth; but

he rights himself at the last moment, and glides away and up again with ease. Occasionally two or three rooks may be seen doing this at once, while the rest whirl about as if possessed; and those that are diving utter a gurgling sound like the usual cawk prolonged – 'caw-wouk'. These antics are believed to foretell rough winds.

The rook, like other broad-winged birds, often makes much leeway in flying, though there be only a moderate wind. The beak points in one direction, in which the bird is apparently proceeding, but if observed closely it will be found that the real course is somewhat sideways. He is making leeway. So it is that a rook which looks as if coming straight towards you – as if he must inevitably go overhead – passes some distance to one side. He appears slow on the wing, as if to go fast required more energy than he possessed, yet he travels over great distances without the least apparent exertion.

When going with the wind he sails high in the air, only flapping his wings sufficiently to maintain balance and steering power. But when working against the wind, if it is a strong gale, his wings are used rapidly, and he comes down near the surface of the ground. He then flies just above the grass, only high enough to escape touching it, and follows the contour of the field. At the hedges he has to rise, and immediately meets the full force of the breeze. It is so powerful sometimes that he cannot overcome it, and his efforts simply lift him in the air, like a kite drawn against the wind. For a few moments he appears stationary, his own impetus and the contending wind balancing each other, and holding him suspended. Then he rises again, but still finding the current too strong, tacks like a ship to port or starboard, and so works aslant into the gale. Shortly afterwards he comes down again, if the field be a large one, and glides forward in the same manner as before, close to the surface. In crossing the lake, too, against the wind, he flies within a few feet of the water.

During such a gale a rook may often be seen struggling to get over a row of trees, and stationary though using his wings vigorously, suspended a little way above the topmost branches. Frequently he has to give up the attempt, turn back, and make a detour.

Though rooks usually go in flocks, individuals sometimes get separated, and may be seen flying alone on the way to rejoin their friends. A flock of rooks, on rising, occasionally divides into two or more parties. Each section wheels off on its own course, while sometimes a small number of those who chance to be near the centre of the original formation seem at a loss which company to follow, and settle down again on the field. So a dozen or more become separated from the crowd, and presently, when they rise, they too divide; three or four fly one way to join one section, and others take another route. Individuals thus find themselves alone; but that causes them no uneasiness, as they have their well-known places of rendezvous, and have only to fly to certain fields to be sure of meeting their friends, or at most to wait about near the nesting trees till the rest come.

It must not, therefore, be supposed that every one flying alone is a crow. Crows are scarce in comparison with rooks. In severe weather a rook will sometimes venture into the courtyard of the farmstead.

Two rooks marked with white resided at the rookery here for several years. One had sufficient white to be distinguished at a distance; the other seemed to have but one or two feathers, which were, however, visible enough when near the bird. As they have not been seen lately, they have probably been shot by someone who thought it clever to destroy anything out of the ordinary. Most large rookeries can either show a rook with white feathers, or have well-authenticated records of their former existence; but though not rare, people naturally like to preserve them when they do occur, and it is extremely annoying to have them wantonly killed.

FIFTEEN

Rooks Returning to Roost

As evening approaches, and the rooks begin to wing their way
homewards, sometimes a great number of them will alight upon the
steep ascent close under the entrenchment on the downs which has been
described, and from whence the wood and beech trees where they sleep can
be seen. They do not seem so much in search of food, of which probably
there is not a great deal to be found in the short, dried-up herbage and
hard soil, as to rest here, halfway home from the arable fields. Sometimes
they wheel and circle in fantastic flight over the very brow of the down,
just above the rampart; occasionally, in the raw cold days of winter, they
perch, moping in disconsolate mood, upon the bare branches of the clumps
of trees on the ridge.

After the nesting time is over and they have got back to their old habits
– which during that period are quite reversed – it is a sight to see from
hence the long black stream in the air steadily flowing onwards to the wood
below. They stretch from here to the roosting-trees, fully a mile and a half
– literally as the crow flies; and backwards in the opposite direction the
line reaches as far as the eye can see. It is safe to estimate that the aerial
army's line of march extends over quite five miles in one unbroken corps.
The breadth they occupy in the atmosphere varies – now twenty yards,
now fifty, now a hundred, on an average, say fifty yards; but rooks do not
fly very close together, like starlings, and the mass, it may be observed, fly
on the same plane. Instead of three or four layers one above the other, the
greater number pass by at the same height from the ground, side by side
on a level, as soldiers would march upon a road: not meaning, of course,
an absolute, but a relative level. This formation is more apparent from an
elevation up among them than from below; and looking along their line
towards the distant wood it is like glancing under a black canopy.

Small outlying parties straggle from the line – now on one side, now on the other; sometimes a few descend to alight on trees in the meadows, where doubtless their nests were situated in the spring. For it is a habit of theirs months after the nesting is over, and also before it begins, to pay a flying visit to the trees in the evening, calling *en route* to see that all is well and to assert possession.

The rustling sound of these thousands upon thousands of wings beating the air with slow steady stroke can hardly be compared to anything else in its weird oppressiveness: it is a little like falling water, but may be best likened, perhaps, to a vast invisible broom sweeping the sky. Every now and then a rook passes with ragged wing – several feathers gone, so that you can see daylight through it; sometimes the feathers are missing from the centre, leaving a great gap, so that it looks as if the bird had a large wing on this side and on the other two narrow ones. There is a rough resemblance between these and the torn sails of some of the old windmills which have become dark in colour from long exposure to the weather, and have been rent by the storms of years. Rooks can fly with gaps of astonishing size in their wings, and do not seem much incommoded by the loss – caused, doubtless, by a charge of shot in the rook shooting, or by the small sharp splinters of flint with which the birdkeepers sometimes load their guns, not being allowed to use shot.

Near their nesting trees their black feathers may be picked up by dozens in the grass; they beat them out occasionally against the small boughs, and sometimes in fighting. If seen from behind, the wings of the rook, as he spreads them and glides, slowly descending, preparatory to alighting, slightly turn up at the edges like the rim of a hat, but much less curved. From a distance as he flies he appears to preserve a level course, neither rising nor falling; but if observed nearer it will be seen that with every stroke of the wings the body is lifted some inches, and sinks as much immediately afterwards.

As the black multitude floats past overhead with deliberate, easy flight, their trumpeters and buglemen, the jackdaws – two or three to every company – utter their curious chuckle; for the jackdaw is a bird which could not keep silence to save his life, but must talk after his fashion, while his grave, solemn companions move slowly onwards, rarely deigning to 'caw' him a reply. But away yonder at the wood, above the great beech trees, where so vast a congregation is gathered together, there is a mighty uproar and commotion: a seething and bubbling of the crowds, now settling on the branches, now rising in sable clouds, each calling to the other with all his might, the whole population delivering its opinions at once.

It is an assemblage of a hundred republics. We know how free States indulge in speech with their parliaments and congresses and senates, their public meetings, and so forth: here are a hundred such nations, all with perfect liberty of tongue, holding forth unsparingly, and in a language which consists of two or three syllables indefinitely repeated. The din is wonderful – each republic as its forces arrive adding to the noise, and for a long time unable to settle upon their trees, but feeling compelled to wheel around and discourse. In spring each tribe has its special district, its own canton and city, in its own trees away in the meadows. Later on they all meet here in the evening. It is a full hour or more before the orations have all been delivered, and even then small bands rush up into the air still dissatisfied.

This great stream of rooks passing over the hills meets another great stream as it approaches the wood, crossing up from the meadows. From the rampart there may be seen, perhaps a mile and a half away, a dim black line crossing at right angles – converging on the wood, which itself stands on the edge of the tableland from which the steeper downs arise. This second army is every whit as numerous, as lengthy, and as regular in its route as the first.

Every morning, from the beech trees where they have slept, safe at that elevation from all the dangers of the night, there set out these two vast expeditionary corps. Regularly, the one flies steadily eastwards over the downs; as regularly the other flies steadily northwards over the vale and meadows. Doubtless in different country districts their habits in this respect vary; but here it is always east and always north. If any leave the wood for the south or the west, as probably they do, they go in small bodies and are quickly lost sight of. The two main divisions sail towards the sunrise and towards the north star.

They preserve their ranks for at least two miles from the wood; and then gradually first one and then another company falls out, and wheeling round, descends upon some favourite field, till by degrees, spreading out like a fan, the army melts away. In the evening, the various companies, which may by that time have worked far to the right or to the left, gradually move into line. By-and-by the vanguard comes sweeping up, and each regiment rises from the meadow or the hill, and takes its accustomed place in the return journey.

So that although if you casually observe a flock of rooks in the daytime they seem to wander hither and thither just as fancy leads, or as they are driven by passers-by, in reality they have all their special haunts; they adhere to certain rules, and even act in concert, thousands upon thousands of them at once, as if in obedience to the word of command, and as if aware

of the precise moment at which to move. They have their laws, from which there is no deviation: they are handed down unaltered from generation to generation. Tradition, indeed, seems to be their main guide, as it is with human tribes. They have their particular feeding grounds; and so you may notice that, comparing ten or a dozen fields, one or two will almost always be found to be frequented by rooks while the rest are vacant.

Here, for instance, is a meadow close to a farmstead – what is usually called the homefield, from its proximity to a house – here day after day rooks alight and spend hours in it, as much at their ease as the nag or the lambs brought up by hand. Another field, at a distance, which to the human eye appears so much more suitable, being retired, quiet, and apparently quite as full of food, is deserted; they scarcely come near it. The homefield itself is not the attraction, because other homefields are not so favoured.

The tenacity with which rooks cling to localities is often illustrated near great cities where buildings have gradually closed in around their favourite haunts. Yet on the small waste spots covered with cinders and dust-heaps, barren and unlovely, the rooks still alight; and you may see them, when driven up from such places, perching on the telegraph wires over the very steam of the locomotives as they puff into the station.

I think that neither considerations of food, water, shelter, nor convenience are always the determining factors in the choice made by birds of the spots they frequent; for I have seen many cases in which all of these were evidently quite put on one side. Birds to ordinary observation seem so unfettered, to live so entirely without rhyme or reason, that it is difficult to convey the idea that the precise contrary is really the case.

Returning to these two great streams of rooks, which pour every evening in converging currents from the north and east upon the wood; why do they do this? Why not go forth to the west, or to the south, where there are hills and meadows and streams in equal number? Why not scatter abroad, and return according to individual caprice? Why, to go still further, do rooks manoeuvre in such immense numbers, and crows fly only in pairs? The simple truth is that birds, like men, have a history. They are unconscious of it, but its accomplished facts affect them still and shape the course of their existence. Without doubt, if we could trace that history back, there are good and sufficient reasons why rooks prefer to fly, in this particular locality, to the east and to the north. Something may perhaps be learnt by examining the routes along which they fly.

The second division – that which goes northwards, after flying little more than a mile in a straight line – passes over Wick Farm, and disperses gradually in the meadows surrounding and extending far below it. The rooks whose nests are placed in the elms of the Warren belong to this division, and, as their trees are the nearest to the great central roosting-place, they are the first to quit the line of march in the morning, descending to feed in the fields around their property. On the other hand, in the evening, as the army streams homewards, they are the last to rise and join the returning host.

So that there are often rooks in and about the Warren later in the evening, after those whose habitations are farther away have gone by, for, having so short a distance to fly, they put off the movement till the last moment. Before watches became so common a possession, the labouring people used, they say, to note the passage overhead of the rooks in the morning in winter as one of their signs of time, so regular was their appearance; and if the fog hid them, the noise from a thousand black wings and throats could not be missed.

If, from the rising ground beyond the Warren or from the downs beyond that, the glance is allowed to travel slowly over the vale northwards, instead of the innumerable meadows which are really there, it will appear to consist of one vast forest. Of the hamlet not far distant there is nothing visible but the white wall of a cottage, perhaps, shining in the sun, or the pale blue smoke curling upwards. This wooded appearance is caused by timber trees standing in the hedgerows, in the copses at the corners of the meadows, and by groups and detached trees in the middle of the fields.

Many hedges are full of elms, some have rows of oaks; some meadows have trees growing so thickly in all four hedges as to seem surrounded by a timber wall; one or two have a number of ancient spreading oaks dotted about in the field itself or standing in rows. But there are not nearly so many trees as there used to be. Numerous hedges have been grubbed to make the fields larger.

Within the last thirty years two large falls of timber have taken place, when the elms especially were thrown wholesale. The old men, however, recall a much greater 'throw', as they term it, of timber, which occurred twice as long ago. Then before that they have a tradition that a still earlier throw took place, when the timber chiefly went to the dockyards for the building of those wooden walls which held the world at bay. These traditions go back, therefore, some eighty or a hundred years. One field in particular is pointed out where stood a double row or avenue of great oaks leading to nothing but a farmstead of the ordinary sort, of which there is not the slightest record that it was ever anything else but a farmhouse. Now avenues of great oaks are not planted to lead to farmsteads. Besides these, it is said, there were oaks in most of the fields – oaks that have long since disappeared, the prevalent tree being elm.

While all these throws of timber have successively taken place, no attempt has been made to fill up the gaps; no planting of acorns, no shielding with rails the young saplings from the ravages of cattle. If a young tree could struggle up, it could; if not it perished. At the last two throws, especially, young trees which ought to have been saved were ruthlessly cut down. Yet even now the place is well timbered; so that it is easy to form some idea of the forest-like appearance it must have presented a hundred years ago, when rows of giant oaks led up to that farmhouse door.

Then there are archaeological reasons, which it would be out of place to mention, why in very ancient days a forest, in all probability, stood hereabouts. It seems reasonable to suppose that in one way or another the regular flight of the second army of rooks passing down into this district was originally attracted by the trees. Three suggestions arise out of the circumstances.

The wood in which both streams of rooks roost at night stands on the last slope of the downs; behind it to the south extend the hills, and the open tilled upland plains; below it northwards are the meadows. It has, therefore, much the appearance of the last surviving remnant of the ancient forest. There has been a wood there time out of mind: there are references to the woods of the locality dating from the sixteenth century. Now if we suppose (and such seems to have been really the case) the unenclosed woodlands below gradually cleared of trees, thereby doubtless destroying many rookeries – the rooks driven away would naturally take refuge in the wood remaining. There the enclosure protected them, and there the trees, being seldom or never cut down, or if cut down felled with judgment and with a view to future timber, grew to great size and in such large groups as they prefer. But as birds are creatures of habit, their descendants in the fiftieth generation would still revisit the old places in the meadows.

Secondly, although so many successive throws of timber thinned out the trees, yet there may still be found more groups and rows of elms and oak in this direction than in any other; that is, a line drawn northwards from the remaining wood passes through a belt of well-timbered country. On either side of this belt there is much less timber; so that the rook that desired to build nests beyond the limits of the enclosed wood still found in the old places the best trees for their purpose. Here may be seen far more rookeries than in any other direction. Hardly a farmhouse lying near this belt has got its rookery, large or small. Once these rookeries were established, an inducement to follow this route would arise in the invariable habit of the birds of visiting their nesting trees even when the actual nesting time is past.

Thirdly, if the inquiry be carried still farther back, it is possible that the line taken by the rooks indicates the line of the first clearings in very early days. The clearing away of trees and underwood, by opening the ground and rendering it accessible, must be very attractive to birds, and rooks are particularly fond of following the plough. Now although the district is at present chiefly meadow land, numbers of these meadows were originally ploughed fields, of which there is evidence in the surface of the fields themselves, where the regular lands and furrows are distinctly visible.

One or all of these suggestions may perhaps account for the course followed by the rooks. In any case it seems natural to look for the reason in the trees. The same idea applies to the other stream of rooks which leaves the wood for the eastward every morning, flying along the downs. In describing the hill district, evidence was given of the existence of woods or forest land upon the downs in the olden time. Detached copses and small woods are still to be found; and it happens that a part of this district, in the line of the eastward flight, belonged to a 'chase', of which several written notices are extant.

The habits of rooks seem more regular in winter than in summer. In winter the flocks going out in the morning or returning in the evening appear to pass nearly at the same hour day after day. But in summer they often stay about late. This last summer I noticed a whole flock, some hundreds in number, remaining out till late – till quite dusk – night after night, and always in the same place. It was an arable field, and there they stood close together on the ground, so close that in spots it was difficult to distinguish individuals. They were silent and still, making no apparent attempt at feeding. The only motion I observed was when a few birds arrived and alighted among them. Where they thus crowded together the earth was literally black.

It was about three-quarters of a mile from their nesting trees, but nesting had been over for more than two months. This particular field had recently been ploughed by steam tackle, and was the only one for a considerable distance that had been ploughed for some time. There they stood motionless, side by side, as if roosting on the ground; possibly certain beetles were numerous just there (for it was noticeable that they chose the same part of the field evening after evening), and came crawling up out of the earth at night.

The jackdaws, which – so soon as the rooks pack after nesting and fly in large flocks – are always with them, may be distinguished by their smaller size and the quicker beats of their wings, even when not uttering their well-known cry. Jackdaws will visit the hencoops if not close to the house, and help themselves to the food meant for the fowls. Poultry are often kept in rickyards, a field or two distant from the homestead, and it is then amusing to watch the impudent attempts of the jackdaws at robbery. Four or five will perch on the post and rails, intent on the tempting morsels: sitting with their heads a little on one side and peering over. Suddenly one thinks he sees an opportunity. Down he hops, and takes a peck, but before he has hardly seized it, a hen darts across, running at him with beak extended like lance in rest. Instantly he is up on the rail again, and the impetus of the hen's charge carries her right under him.

Then, while her back is turned, down hops a second and helps himself freely. Out rushes another hen, and up goes the jackdaw. A pause ensues for a few minutes: presently a third black rascal dashes right into the midst of the fowls, picks up a morsel, and rises again before they can attack him. The way in which the jackdaw dodges the hens, though alighting among them, and as it were for the moment surrounded, is very clever; and it is laughable to see the cool impudence with which he perches again on the rail and looks down demurely, not a whit abashed, on the feathered housewife he has just been doing his best to rob.

SIXTEEN

Notes on Birds

THE NIGHTINGALE is one of the birds whose habit of returning every year to the same spot can hardly be overlooked by anyone. Hawthorn and hazel are supposed to attract them: I doubt it strongly. If there is a hawthorn bush near their favourite resting place they will frequent it by choice, but of itself it will not bring nightingales. They seem to fix upon localities in the most capricious manner. In this particular district they are moderately plentiful; yet in the whole of a large parish (some five miles across) they are only found in one place. The wood, which is the roosting-place of all the rooks, large as it is, has but one haunt of the nightingale. Just in one special spot they may be heard, and nowhere else. But having selected a locality, they come back to it as regularly as the swallows.

In another county in the same latitude there is a small copse of birch which borders a much-frequented road. Here the stream of vehicles and passengers is nearly continuous; and the birch copse abounds with nightingales in the spring. On one fine morning I counted eight birds singing at once. The young birds seemed afterwards as numerous as the sparrows. Never, in the wildest district I have ever visited, have I seen so many. They had become so accustomed to passers-by that they took no notice unless purposely disturbed. Several times I stood under an oak bough that projected across the sward by the roadside, with a nightingale perched on it overhead straining his throat. The bough was some twelve feet high, and in full view of everyone. This road was constructed about a hundred years ago; and it would be interesting to learn if a country lane preceded it, well sheltered on both sides by thick hedges. Birds are fond of such places, and, having once formed the habit of coming there, would continue to do so after the highway was laid down.

It has been stated that the flocks of chaffinches which may be seen in winter consist entirely of females. Male chaffinches are rarely seen: they have migrated, or in some other manner disappeared. Yet so soon as the spring comes on the males make their presence known by calling their defiant notes from every elm along the road. Last spring I fell into conversation with a fowler. He had a cock chaffinch in a cage covered with a black cloth, except on one side. The cage was placed on the sward beside the road, and near it a stuffed cock bird stood on the grass. Two pieces of whalebone smeared with birdlime formed a pointed arch over the stuffed chaffinch. The live decoy bird in the cage from time to time uttered a few notes, which were immediately answered by a wild bird in the elms overhead. These notes are a challenge; and the bird in the tree supposes them to proceed from the stuffed bird in the grass, and descends to fight him, when, as the deceived bird alights, his wings or feet come in contact with the whalebone – sometimes he perches on it – and the lime holds him fast.

At that season (March) the cock birds have an irresistible inclination to do battle; they are ceaselessly challenging each other, and the fowler takes advantage of it to snare them. Now this man said that these chaffinches sold for 6s. the dozen, and that when the birds were 'on', as he called it, he could catch five dozen a day. In a walk of four or five miles I passed half a dozen such fellows, with cages and stuffed chaffinches. This alone proves that cock chaffinches are very numerous in spring. Where, then, are they in winter, if the flocks of chaffinches at that period consist almost exclusively of female birds? Probably they fly in small bodies of three or four, or singly, and so escape observation. But this division of the sexes presents a curious resemblance to the social customs discovered amongst certain savages. During the winter the birds separate, and the females 'pack'. In the spring the males appear, and, after a period of fighting for the mastery, pair, and the nests are built. After the young are reared, song ceases, and the old haunts are deserted. This summer I was much struck with this partial migration, perhaps the more so because observed in a fresh locality.

During the spring and summer I daily followed a road for some three miles which I had found to pass through a district much frequented by birds. The birch coppice so favoured by nightingales was that way: and, by-the-by, the wrynecks were almost equally numerous; and the question has occurred to me whether these birds are companions, in a sense, of the nightingale, having noticed them in other places to be much together. All spring and summer the hedges, coppices, brakes, thickets, furze lands, and cornfields abounded with bird life. About the middle of August there was a notable decrease. Early in September the places previously so populous seemed almost deserted; by the middle of the month quite deserted.

There were no chaffinches in the elms or in the road, and scarcely a sparrow; not a yellowhammer on the hedge by the cornfield; only a very few greenfinches; not a single bullfinch or goldfinch. Blackbirds, thrushes, and robins alone remained. The way to find what birds are about is to watch one of their favourite drinking and bathing places; then it is easy to see which are absent. Where had all those birds gone? In the middle of the fields of stubble there were flocks of sparrows – innumerable sparrows – and some finches, but not, apparently, enough to account for all that had left the hedges and trees. That may be explained by their being scattered over so many broad acres – miles of arable land being open to them.

But the migration from the hedgerows was very marked. They became quite empty and silent about the middle of September. This state of things continued for little more than a week – meaning the absolute silence – then a bird or two appeared in places at long intervals. They now came back rapidly, till, on the 28th, the 'fink, chink' of the finches sounded almost as merrily as before. The greenfinches flew from tree to tree in parties of four, six, or more, calling to each other in their happy confidential way. On that day the trees and hedges seemed to become quite populous again with finches. The sparrows, too, were busy in the roads once more. For a week previously every now and then a single lark might be heard singing for a few minutes; they had been silent before. On the 28th half a dozen could be heard singing at once, and now and then a couple might be seen chasing each other as if full of gaiety. It was indeed almost like a second spring: at the same time a few buttercups bloomed, to add to the illusion.

This migration of the finches from the hedgerows out into the fields, and their coming back, is very striking. It may possibly be connected with the phenomenon of 'packing'; for they seem to go away by twos and threes, to disappear gradually, but to return almost all at once, and in parties or flocks. The number in the flocks varies a great deal: it is a common opinion that it depends on the weather, and that in hard winters, when the cold is severe and prolonged, the flocks are much larger. Wood pigeons are seldom, it is said, seen in great flocks till the winter is advanced.

Has the date of the harvest any influence upon the migration of birds? The harvest in some counties is, of course, much earlier than in others – a fact of which the itinerant labourer takes advantage, following the wave of ripening grass and corn. By the time they have mown the grass or reaped the wheat, as the case may be, in one county, the crops are ripe in another, to which they then wend their way.

One of the very earliest counties, perhaps, is Surrey. The white bloom of the blackthorn seems to show there a full fortnight earlier than it does on the same line of latitude not many miles farther west. The almond trees exhibit their lovely pink blossom; the pears bloom, and presently the hawthorn comes out into full leaf, when a degree of longitude to the west the hedges are bare and only just showing a bud. Various causes probably contribute to this – difference of elevation, difference of soil, and so forth. Now the spring visitors – as the cuckoo, the swallow, and wryneck – appear in Surrey considerably sooner than they do farther west. The cuckoo is sometimes a full week earlier. It would seem natural to suppose that the more forward state of vegetation in that county has something to do with the earlier appearance of the bird. But I should hesitate to attribute it entirely to that cause, for it sometimes happens that birds act in direct opposition to what we should consider the most eligible course.

For instance, the redwing is one of our most prominent winter visitors. Flocks of redwings and fieldfares are commonly seen during the end of the season. They come as winter approaches, they leave as it begins to grow warm. In every sense they are birds of passage; any ploughboy will tell you so. (By-the-by, the ploughboys call the fieldfares 'velts'. Is not velt a Northern word for field?) But one spring – it was rapidly verging on summer – I was struck day after day by hearing a loud, sweet, but unfamiliar note in a certain field. Fancying that most bird notes were known to me, this new song naturally arrested my attention. In a little while I succeeded in tracing it to an oak tree. I got under the oak tree, and there on a bough was a redwing singing with all his might. It should be remarked that neither redwing nor fieldfare sings during the winter; they of course have their call and cry of alarm, but by no stretch of courtesy could it be called a song. But this redwing was singing – sweet and very loud, far louder than the old familiar notes of the thrush. The note rang out clear and high, and somehow sounded strangely unfamiliar among English meadows and English oaks.

Then, looking farther and watching about the hedges there, I soon found that the bird was not alone – there were three or four pairs of redwings in close neighbourhood, all evidently bent upon remaining to breed. To make quite sure, I shot one. Afterwards I found a nest and had the pleasure of seeing the young birds come to maturity and fly.

Nothing could be more thoroughly opposed to the usual habits of the bird. There may be other instances recorded, but what one sees oneself leaves so much deeper an impression. The summer that followed was a very fine one. It is instances like this that make one hesitate to dogmatise too much as to the why and wherefore of bird-ways. Yet it is just the speculation as to that why and wherefore which increases the pleasure of observing them.

Then there is the corncrake, of whose curious tricks in the mowing-grass I have already written. The crake's rules of migration are not easily reconciled with any theory I have ever heard of. In the particular locality which has been described the crakes come early, they enter the mowing-grass and remain there till after it is cut; immediately afterwards they are heard in the corn. Presently they are silent and supposed to be gone; but I have heard of their being shot on the opening of the shooting season on the uplands. The cry of the crake in that locality is so common and so continuous as to form one of the most striking features of the spring: the farmers listen for them and note their first arrival, just as for the cuckoo – which, it may be observed in passing, even in England keeps time with the young figs.

But when I had occasion to pass a spring in Surrey the first thing I noticed was the rarity of the crakes; I heard one or two at most, and that only for a short time. Long before the grass was mown they were gone – doubtless northwards, having only called in passing. I am told they call again in coming back, and are occasionally shot in September. But the next spring, chancing again to be in Surrey at that season, though constantly about out of doors, I never heard a crake but once – one single call – and even then was not quite sure of it. I am told, again, that there are parts of the county where they are more numerous; they were certainly scarce those two seasons in that locality. Now here we have an instance in direct contradiction to the suggestion that the early state of vegetation is attractive to our spring visitors. The crakes appeared to come earlier, in larger numbers, and to be more contented and make a longer stay in the colder county than in the warm one.

The packing of birds is very interesting, and no thoroughly satisfactory explanation of it, that I am aware of, has ever been discovered. It is one of the most prominent facts in their history. It is not for warmth, because they pack long before it is cold. This summer I saw large flocks of starlings flying to their favourite firs to roost on the evening of the 19th of June. The cuckoo was singing on the 17th, two days before.

It would be interesting to know, too, whether birds are really as free in the choice of their mates in spring as at first sight appears. They return to the same places, the same favourite hedge, and even the same tree. Now, when the flocks split up into sections as the spring draws near, each section or party seems to revisit the hedge from which they departed last autumn. Do they, then, intermarry year after year? and is that the reason why they return to the same locality? The fact of a pair building by chance in a certain hedge is hardly enough to account for the yearly return of birds to the spot. It seems more like the return of a tribe or *gens* to its own special locality. The members of such a *gens* must in that case be closely related. As it is not possible to identify individual birds, the difficulty of arriving at a clear understanding is great.

Why, again, do not robins pack? Why do not blackbirds and thrushes go in flocks? They never merge their individuality all the year round. Even herons, though they fish separately, are gregarious in building, and also often in a sense pack during the day, standing together on a spit or sandbank. Rooks, starlings, wood pigeons, fieldfares, and redwings may be seen in winter all feeding in the same field, and all in large flocks.

Some evidence of a supposed tendency to intermarry among birds may perhaps be deduced from the practice of the long-tailed titmouse. This species builds a nest exactly like a hut, roof included, and in it several birds lay their eggs: as many as twenty eggs are sometimes found; fourteen is a common number. Here there is not only the closest relationship, but a system of community. This tit has a way sometimes of puffing up its feathers – they are fluffy, and in that state look like fur – and uttering a curious sound much resembling the squeak of a mouse; hence, perhaps, the affix 'mouse' to its name.

The tomtit also packs, and flies in small parties almost all the year round. They remain in such parties until the very time of nesting. On March 24 last, while watching the approach of a snowstorm, I noticed that a tall birch tree – whose long, slender, weeping branches showed distinctly against the dark cloud – seemed to have fruit hanging at the end of several of the boughs. On going near I counted six tomtits, as busy as they could be, pendent from as many tiny drooping boughs, as if at the end of a string, and swinging to and fro as the rude blast struck the tree. The six in a few minutes increased to eight, then to nine, then to twelve, and at last there were fourteen altogether, all dependent from the very tiniest drooping boughs, all swinging to and fro as the snowflakes came silently floating by, and all chuckling and calling to each other. The ruder the blast and the more they swung – head downward – the merrier they seemed, busily picking away at the young buds. Some of them remained in the tree more than an hour.

Peewits or lapwings not only pack in the winter, but may almost be said to pass the nesting time together. There are two favourite localities in the district, which has been more particularly described, much frequented by these birds. One is among some water meadows, where the grass is long earlier in the spring than elsewhere: there the first bennet pushes up its

green staff – country people always note the appearance of the first bennet – and the first cuckoo flower opens. Several nests are made here on the ground, in comparatively close contiguity.

Upon approaching, the old bird flies up, circles round, and comes so near as almost to be within reach, whistling 'pee-wit, pee-wit', over your head. He seems to tumble in the air as if wounded and scarcely able to fly; and those who are not aware of his intention may be tempted to pursue, thinking to catch him. But so soon as you are leaving the nest behind he mounts higher, and wheels off to a distant corner of the field, uttering an ironical 'pee-wit' as he goes. If you neglect his invitation to catch him if you can, and search for the nest or stand still, he gets greatly excited and comes much closer, and in a few minutes is joined by his mate, who also circles round; while several of their friends fly at a safer distance, whistling in sympathy.

Then you have a good opportunity of observing the peculiar motion of their wings, which seem to strike simply downwards and not also backwards, as with other birds; it is a quick jerking movement, the wing giving the impression of pausing the tenth of a second at the finish of the stroke before it is lifted again. If you pass on a short distance and make no effort to find the nest, they recover confidence and descend. When the peewit alights he runs along a few yards rapidly, as if carried by the impetus. He is a handsome bird, with a well-marked crest.

The other locality to which I have referred was a wide open field full of anthills. There must have been eight or ten acres of these hills. They rose about eighteen inches or two feet, of a conical shape, and overgrown by turf, like thousands of miniature extinct volcanoes. They were so near together that it was easy to pass twenty or thirty yards, without once touching the proper surface of the ground, by springing from one anthill to the other. Thick bunches of rushes grew between, and innumerable thistles flourished, and here and there scattered hawthorn bushes stood. It was a favourite place with the finches; the hawthorn bushes always had nests in them. Thyme grew luxuriantly on the ground between the nests and on the anthills. Wild thyme and ants are often found together as on the downs. How many millions of ants must have been needed to raise these hillocks!

and what still more incalculable numbers must have lived in them! A wilder spot could scarcely have been imagined, though between rich meadow and ploughed lands.

There was always a covey of partridge about the field, but they could not have had such a feast of eggs as would naturally be supposed, because in the course of time a crust of turf had grown over the anthills. The temporary hills of loose earth thrown up every summer by the sides of the fields, where they can lay bare a whole nest with two or three scratches, must afford much more food. Had it been otherwise all the partridges in the neighbourhood would have gathered together here; but there never seemed more than one or two coveys about.

The peewits had nests year after year in this place, and even when the nesting time was over a few might often be seen. The land for agricultural purposes was almost valueless, there being so little herbage upon which cattle could graze, and no possibility of mowing any; so in the end gangs of labourers were set to work and the anthills levelled, and, indeed, bodily removed. Thus this last piece of waste land was brought into use.

Upon the downs there is a place haunted by some few peewits. In the colder months they assemble in flocks, and visit the arable land where it is of a poor character, or where there are signs of peat in the soil. By the shores of the lake they may, too, be often seen. I have counted sixty in one flock, and have seen flocks so numerous as to be unable to count them accurately; that, of course, was exceptional, but they are by no means uncommon birds in this district. In others it seems quite a rare thing to see a lapwing.

They often appear to fly for a length of time together for the mere pleasure of flying. They rise without the slightest cause of alarm, and sail about to and fro over the same field for half an hour, then settle and feed again, and presently take wing and repeat the whirling about overhead. Solitary peewits will do the same thing; you would imagine they were going off at a great pace, instead of which back they come in a minute or two. Other birds fly for a purpose: the peewit seems to find enjoyment beating to and fro in the air.

Crows frequently build in oaks, and unless they are driven away by shot will return to the same neighbourhood the following year. They appear to prefer places near water, and long after the nesting time is past will visit the spot. Small birds will sometimes angrily pursue them through the air, as they will hawks. As autumn approaches the swallows congregate on warm afternoons on church steeples; they may be seen whirling round and round in large flocks, and presently settling. I saw a crow go past a steeple where there was a crowd of swallows, when immediately the whole flock took wing, and circled about the crow, following him for some distance. He made an awkward attempt once to get at some of them, but their swiftness of wing took them far out of his reach. Crows make no friends; rooks, on the contrary, make many, and are often accompanied by several other species of birds. A certain friendliness, too, seems to exist between sparrows, chaffinches, and greenfinches, which are often found together.

Some fields are divided into two by a long line of posts and rails, which in time become grey from the lichen growing on the wood. The cuckoos in spring seem to like resting on such rails better than the hedges; and when they are courting, two, or even three, may be sometimes seen on them together. Presently they fly, and are lost sight of behind the trees: but one or other is nearly sure to come back to the rails again after a while. Cuckoos perch frequently, too, on those solitary up-right stones which here and there stand in the midst of the fields. This habit of theirs is quoted by some of the old folks as an additional proof that the cuckoo is only a hawk changed for the time, and unable to forget his old habits, hawks (and owls) perching often on poles, or anything upright and detached.

The cuckoo flies so much like the hawk, and so resembles it, as at the first glance to be barely distinguishable; but on watching more closely it will be seen that the cuckoo flies straight and level, with a gentle fluttering of the wings, which never seem to come forward, so that in outline he resembles a crescent, the convex side in front. His tail appears longer in proportion, and more pointed; his flight is like that of a very large swallow flying straight. The cuckoo's cry can perhaps be heard farther than the call of any other bird. The heron's power of voice comes nearest: he sails at a great height, and his 'quaaack', drawn out into a harsh screech, may be heard at a long distance. But then he has the advantage of elevation; the cuckoo never rises above the tops of the elms.

Yellowhammers have a habit of sitting on a rail or bough with their shoulders humped, so that they seem to have no neck. In that attitude they will remain a long time, uttering their monotonous chant; most other birds stretch themselves and stand upright to sing. The great docks that grow beside the ditches are visited by the tomtits, who perch on them – the stalk of the dock is strong and supports so light a weight easily. Sparrows may sometimes be seen in July hawking in the air just above the sward by the roadside – hovering like the kestrel, a foot or so high, and then suddenly dropping like stones; they are then so absorbed that they will scarcely fly away at your approach. At the same time a rather long red fly is abundant in the grass, and may be the attraction. The swift's long narrow wings shut behind him as if with a sharp snip, cutting the air like shears; and then, holding them extended, he glides like a quoit.

In the old days men used to be on the watch about the time of the great race meetings, in order to shoot at every pigeon that went past, in hope of finding a message attached to the bird, and so getting the advantage of early intelligence. In one such case I heard of, the pigeon had the name of the winner and was shot on a tree where it had alighted weary from want of food or uncertain as to its course.

The golden-crested wren – smallest of the birds – scarcely ever leaves the shelter of the hedges and trees. The crest or top-knot is not exactly golden, but rather orange; and as the body of the tiny creature is dusky in hue, the bright colour on its head shines like flame in contrast. By this ruddy lamp upon its head the wren may be discovered hidden deep in the intricate mazes of the thorn bushes, where otherwise it would be difficult to find. These wrens are usually in pairs; I have seldom seen one by itself. They are not rare, and yet are comparatively little seen, and must, I think, travel a good deal. All the same, they have their favourite places; there was one hedge where, if the bird was anywhere in the neighbourhood, I could feel sure of finding him. It was very thick, and entirely of hawthorn and blackthorn, and divided two water meadows.

SEVENTEEN

Notes on the Year

THERE ARE FEW HEDGES so thick that in January it is not possible to see through them, frost and wind having brought down the leaves. The nettles, however, and coarse grasses, dry brown stems of dead plants, rushes, and moss, still in some sense cover the earth of the mound, and among them the rabbits sit out in their forms. Looking for these with gun and spaniel, when the damp mist of the morning has cleared, one sign – one promise – of the warm days to come may chance to be found. Though the sky be gloomy, the hedge bare, and the trees gaunt, among the bushes a solitary green leaf has already put forth. It is on the stalk of the woodbine which climbs up the hawthorn, and is the first in the new year – in the very darkest and blackest days – to show that life is stirring. As it is the first to show a leaf, so, too, it is one of the latest to yield to the advancing cold, and even then its bright red berries leave a speck of colour; and its bloom, in beauty of form, hue, and fragrance, is not easily surpassed.

While the hedges are so bare the rabbits are unmercifully ferreted, for they will before long begin to breed. On the milder mornings the thrushes are singing sweetly. Clouds of tiny gnats circle in the sheltered places near houses or thatch. In February 'fillditch', as the old folk call it, on account of the rains, although nominally in the midst of the winter quarter, there is a distinct step forward. If the clouds break and the wind is still, the beams of the sun on the southern side of the wall become pleasantly genial. In the third week they bring forth the yellow butterfly, fluttering gaily over the furze; while the larks on a sunny day, chasing each other over the ploughed fields, make even the brown clods of earth seem instinct with awakening life. The pairing off of the birds is now apparent in every hedge, and at the same time on the mounds, and under sheltering bushes and trees a deeper green begins to show as the plants push up.

The blackthorn is perhaps the first conspicuous flower; but in date it seems to vary much. On the 22nd of February, 1877, there were boughs of blackthorn in full bloom in Surrey, and elder trees in leaf; nearly three weeks before that, at the beginning of the month, there were hawthorn branches in full leaf in a sheltered nook in Kent. A degree farther west, on the contrary, the hawthorn did not show a leaf for some time after the blackthorn had bloomed in Surrey. The farmers say that the grass which comes on rapidly in the latter days of February and early days of March, 'many weathers' (in their phrase), often 'goes back' later in the season, and loses its former progress.

Lady Day (old style) forms with Michaelmas the two eras, as it were, of the year. The first marks the departure of the winter birds and the coming of the spring visitors; the second, in reverse order, marks the departure of the summer birds and the appearance of the vanguard of the winter ones. In the ten days or fortnight succeeding Lady Day – say from the 6th of April to the 20th – great changes take place in the fauna and flora; or, rather, those changes which have long been slowly maturing become visible. The nightingales arrive and sing, and with them the white butterfly appears. The swallow comes, and the wind anemone blooms in the copse. Finally the cuckoo cries, and at the same time the pale lilac cuckoo flower shows in the moist places of the mead.

The exact dates, of course, vary with the character of the season and the locality; but, speaking generally, you should begin to keep a keen lookout for these signs of spring about old Lady Day. In the spring of last year, in a warm district, the nightingale sang on the 12th of April, a swallow appeared on the 13th, and the note of the cuckoo was heard on the 15th. No great reliance should be put upon precise dates, because in the first place they vary annually, and in the next an observer can, in astronomical language, only sweep a limited area, and that but imperfectly; so that it is very likely some ploughboy who thinks nothing of it – except to immediately imitate it – hears the cuckoo forty-eight hours before those who have been listening most carefully. So that these dates are not given because they are of any intrinsic value, but simply for illustration. On the 14th of April (the same spring) the fieldfares and redwings were passing over swiftly in small parties – or, rather, in a long flock scattered by the march – towards the North Sea and their summer home in Norway. The winter birds, and the distinctly spring and summer birds, as it were, crossed each other and were visible together, their times of arrival and departure overlapping.

As the sap rises in plants and trees, so a new life seems to flow through the veins of bird and animal. The flood tide of life rises to its height, and after remaining there some time, gradually ebbs. Early in August the leaves of the limes begin to fade, and a few shortly afterwards fall; the silver birch had spots of a pale lemon among its foliage this year on August 13. The brake fern, soon after it has attained its full growth, begins to turn yellow in places. There is a silence in the hedges and copses, and an apparent absence of birds. But about Michaelmas (between the new and old styles) there is a marked change. It is not that anything particular happens upon any precise day, but it is a date around which, just before and after, events seem to group themselves.

Towards the latter part of September the geometrical spiders become conspicuous, spinning their webs on every bush. Some of these attain an enormous size, and, being so large, it is easier to watch their mode of procedure. When a fly becomes entangled, the spider seizes it by the poll, at the back of the head, and holds it for a short time till it dies. Then he rapidly puts a small quantity of web round it, and next carries it to the centre of the web. There, taking the dead fly on his feet – much as the juggler plays with a ball upon his toes – the spider rolls it round and round, enveloping it in a cocoon of web, and finally hangs up his game head uppermost, and resumes his own position head downwards. Another spider wraps his prey in a cocoon by spinning himself and the fly together round and round. At the end of September or beginning of October acres of furze may be seen covered with web in the morning, when the dew deposited upon it renders it visible. As the sun dries up the dew the web is no longer seen.

On September 21 of last year the rooks were soaring and diving; they continued to do this several days in succession. I should like to say again that I attach no importance to these dates, but give them for illustration;

these, too, were taken in a warm district. Rooks usually soar a good deal about the time of the equinox. On September 29 the heaths and furze were white with the spiders' webs. September 27, larks singing joyously. October 2, a few grasshoppers still calling in the grass – heard one or two three or four days later. October 4, the ivy in full flower. October 7, the thrushes singing again in the morning. October 6 and 7, pheasants roaming in the hedges for acorns. October 13, a dragonfly – large and green – hawking to and fro on the sunny side of hedge. October 15, the first redwing. During the latter part of September and beginning of October, frogs croaking in the ivy.

Now, these dates would vary greatly in different localities, but they show, clearer than a mere assertion, that about that time there is a movement in nature. The croaking of frogs, the singing of larks and thrushes, are distinctly suggestive of spring (the weather, too, was warm and showery, with intervals of bright sunshine); the grasshopper and dragonfly were characteristic of summer, and there were a few swallows still flying about; the pheasants and the acorns, and the puff balls, full of minute powder rising in clouds if struck, spoke of autumn; and, finally, the first redwing indicated winter; so that all the seasons were represented together in about the space of a fortnight. I do not know any other period of the year which exhibits so remarkable an assemblage of the representative features of the four quarters; an artist might design an emblematic study upon it, say, for a tessellated pavement.

In the early summer the lime trees flower, and are then visited by busy swarms of bees, causing a hum in the air overhead. So, in like manner, on October 16, I passed under an old oak almost hidden by ivy, and paused to listen to the loud hum made by the insects that came to the ivy blossom. They were principally bees, wasps, large black flies, and tiny gnats. Suddenly a wasp attacked one of the largest of the flies, and the two fell down on a bush, where they brought up on a leaf.

The fly was very large, of a square build, and wrestled with its assailant vigorously. But in a few seconds, the wasp, getting the mastery, brought his tail round, and stung the fly twice, thrice, in rapid succession in the abdomen, and then held tight. Almost immediately the fly grew feeble; then the wasp snipped off its proboscis, and next the legs. Then he seized the fly just behind the head, and bit off pieces of the wings; these, the proboscis, and the legs dropped to the ground. The full purpose of the wasp is not easily described; he stung and snipped and bit and reduced his prey to utter helplessness, without the pause of a second.

So eager was he that while cutting the wings to pieces he fell off the leaf, but clung tight to the fly, and although it was nearly as big as himself, carried it easily to another leaf. There he rolled the fly round, snipped off the head, which dropped, and devoured the internal part; but slipped again and recovered himself on a third leaf, and picked the remaining small portion. What had been a great insect had almost disappeared in a few minutes.

After the arrival of the fieldfares the days seem to rapidly shorten, till towards the end of December the cocks, reversing their usual practice, crow in the evening, hours before midnight. The cockcrow is usually associated with the dawn, and the change of habit just when the nights are longest is interesting.

Birds have a feng shui of their own – an unwritten and occult science of the healthy and unhealthy places of residence – and seem to select localities in accordance with the laws of this magical interpretation of nature. The sparrows, by preference, choose the southern side of a house for their nests. This is very noticeable on old thatched houses, where one slope of the roof happens to face the north, and another the south. On the north side the thatch has been known to last thirty years without renewal – it decays so slowly. The moss, however, grows thickly on that side, and if not removed would completely cover it. Moss prefers the shade; and so in the woodlands, the meadows on the north, or shady side of the copses are often quite overgrown with moss, which is pleasant to walk on, but destroys the herbage. But on the south side of the roof, the rain coming from that quarter, the wind and sun cause the thatch to rapidly deteriorate, so that it requires to be constantly repaired.

Now, instead of working their holes into the northern slope sheltered from wind and rain, nine out of ten of the sparrows make their nests on the south, and, of course, by pulling out the straw still further assist the decay of the thatch there. The influence of light seems to be traceable in this; and it does occur to me that other birds that use trees and bushes for their nests may be guided in their selection by some similar rule. The trees and bushes they select to us look much the same as others; but the birds may none the less have some reasons of their own. And as certain localities, as previously observed, are great favourites with them and others are deserted, possibly feng shui may have something to do with that also.

The nomadic tribes that live in tents, and wander over thousands of miles in the East, at first sight seem to roam aimlessly, or to be determined simply by considerations of water and pasture. But those who have lived with and studied them say that, though they have no maps, each tribe, and even each particular family, has its own special route and special camping ground. If these routes were mapped out, they would present an interlaced pattern of lines crossing and recrossing without any appreciable order; yet one family never interferes with another family. This statement seems to me to be most interesting if compared with the habits of birds that roam hither and thither apparently without order or method, that come back in the spring to particular places, and depart again after their young are reared. Though to us they wander aimlessly, it is possible that from their point of view they may be following strictly prescribed routes sanctioned by immemorial custom.

And so itinerant labourers move about. In the early spring they go to the uplands, where there are many thousand acres of arable land, for the hoeing. Then comes a short space of employment – haymaking in the water meadows that follow the course of the rivers there, and which are cut very early. Next, they return down into the vale, where the haymaking has then commenced. Just before it begins the Irish arrive in small parties, coming all the way from their native land to gather the high wages paid during the English harvest. They show a pleasing attachment to the employer who has once given them work and treated them with a little kindness. To him they go first; and thus it often happens that the same band of Irish return to the same farm year after year as regularly as the cuckoo. They lodge in an open shed, making a fire in the corner of the hedge where it is sheltered. They are industrious, work well, drink little, and bear generally a good character.

After the haymaking in the vale is finished, the itinerant families turn towards the lighter soils, where the corn crops are fast ripening, and soon leave the scene of their former labours fifty miles behind them. A few, perhaps, straggle back in time to assist in the latter part of the corn harvest on the heavy lands, if it has been delayed by the weather. The physicians say that change of air is essential to health: the migration of birds may not be without its effect upon their lives, quite apart from the search for food alone.

The dry walls which sometimes enclose cornfields (built of flat stones) are favourite places with many birds. The yellowhammers often alight on them, so do the finches and larks; for the coarse mortar laid on the top decays and is overgrown with mosses, so that it loses the hard appearance of a wall. When the sparrow who has waited till you are close to him suddenly starts, his wings, beating the air, make a sound like the string of a bow pulled and released – to try it without an arrow.

The dexterous way in which a bird helps itself to thistledown is interesting to watch. The thistle has no branch on which he can perch; he must take it on the wing. He flies straight to the head of the thistle, stoops as it were, seizes the down, and passes on with it in the bill to the nearest bough – much in the same way as some tribes of horsemen are related to pick up a lance from the ground whilst going at full speed.

Many birds twirl their 'r's'; others lisp, as the nightingale, and instead of 'sweet' say 'thweet, thweet'. The finches call to each other, 'Kywee, kywee – tweo – thweet', which, whatever may be its true translation, has a peculiarly soothing effect on the ear. Swifts usually fly at a great height, and, being scattered in the atmosphere, do not appear numerous; but sometimes during a stiff gale they descend and concentrate over an open field, there wheeling round and to and fro only just above the grass. Then the ground looks quite black with them as they dart over it: they exhibit no fear, but if you stand in the midst come all round you so close that they might be knocked down with a walking stick. In the air they do not look large, but when so near as this they seem to be of considerable size. The appearance of hundreds of these jet black, long-winged birds, flying with marvellous rapidity, and threading an inextricable maze almost, as it were, under foot, is very striking.

The proverbial present of a white elephant is paralleled in bird life by the gift of the cuckoo's egg. The bird whose nest is chosen never deserts the strange changeling, but seems to feel feeding the young cuckoo to be a sacred duty, and sees its own young ejected and perishing without apparent concern. My attention was called one spring to a robin's nest made in a stubble rick; there chanced to be a slight hollow in the side of the rick, and this had been enlarged. A cuckoo laid her egg in the nest, and as it happened to be near some cowsheds it was found and watched. When the young bird began to get fledged some sticks were inserted in the rick so as to form a cage, that it might not escape, and there the cuckoo grew to maturity and to full feather.

All the while the labour undergone by the robins in supplying the wide throat of the cuckoo with food was incredible. It was only necessary to wait a very few minutes before one or other came, but the voracious creature seemed never satisfied; he was bigger than both his foster-parents put together, and they waited on him like slaves. It was really distressing to see their unrewarded toil. Now, no argument will ever convince me that the robin or the wagtail, or any other bird in whose nest the cuckoo lays its egg, can ever confound the intruding progeny with its own offspring. Irrespective of size, the plumage is so different; and there is another reason why they must know the two apart: the cuckoo as he grows larger begins to resemble the hawk, of which all birds are well known to feel the greatest terror. They will pursue a cuckoo exactly as they will a hawk.

I will not say that is because they mistake it for a hawk, for the longer I observe the more I am convinced that birds and animals often act from causes quite distinct from those which at first sight account for their motions. But there is no doubt of the lesser birds chasing the cuckoo. Are they endeavouring to drive her away so that she may not lay her egg in either of their nests? In any case it is clear that birds do recognise the cuckoo as something distinct from themselves, and therefore I will never believe that the foster-parent for a moment supposes the young cuckoo to be its own offspring.

To our eyes one young robin (meaning out of the nest – on the hedge) is almost identical with another young robin; to our ears the querulous cry of one for food is confusingly like that of another: yet the various parent birds easily distinguish, recognise, and feed their own young. Then to suppose that, with such powers of observation – with the keenness of vision that can detect an insect or a worm moving in the grass from a branch twenty feet or more above it, and detect it while to all appearance engaged in watching your approach – to suppose that the robin does not know that the cuckoo is not of its order is past credit. The robin is much too intelligent. Why, then, does he feed the intruder? There is something here approaching the sentiment of humanity, as we should call it, towards the fellow creature.

The cuckoo remained in the cage for some time after it had attained sufficient size to shift for itself, but the robins did not desert it: they clearly understood that while thus confined it had no power of obtaining food and must starve. Unfortunately, a cat at last discovered the cuckoo, which was found on the ground dead but not eaten. The robins came to the spot afterwards – not with food, but as if they missed their charge.

The easy explanation of a blind instinct is not satisfactory to me. On the other hand, the doctrine of heredity hardly explains the facts, because how few birds' ancestors can have had experience in cuckoo-rearing? There is no analogy with the cases of goats and other animals suckling strange species; because in those instances there is the motive – at all events in the beginning – of relief from the painful pressure of the milk. But the robins had no such motive: all their interests were to get rid of their visitor. May we not suppose, then, that what was begun through the operation of hereditary instinct, i.e. the feeding of the cuckoo, while still small and before the young robins had been ejected, was continued from an affection that gradually grew up for the helpless intruder? Higher sentiments than those usually attributed to the birds and beasts of the field may, I think, be

traced in some of their actions.

To the number of those birds whose call is more or less apparently ventriloquial the partridge may be added; for when they are assembling in the evening at the roosting-place their calls in the stubble often sound some way to the right or left of the real position of the bird, which appears emerging from the turnips ten or fifteen yards farther up than judged by the ear. It is not really ventriloquial, but caused by the rapid movements and by the circumstance of the bird while out of sight.

We constantly hear that the area of pasture in England is extending, and gradually overlapping arable lands; and the question suggests itself whether this, if it continues, will not have some effect upon bird and animal life by favouring those that like grass lands and diminishing those that prefer the ploughed. On and near ploughed lands modern agriculture endeavours to cut down trees and covers and grub up hedges, not only on account of their shade and the injury done by their roots, but because they are supposed to shelter sparrows and other birds. But pasture and meadow are favourable to hedges, trees, and covers: wherever there is much grass there is generally plenty of wood; and this again – if hedges and small covers extend in a corresponding degree with pasture – may affect bird life.

A young dog may be taught to hunt almost anything. Young pointers will point birds' nests in hedges or trees, and discover them quicker than any lad. If a dog is properly trained, of course this is not allowed; but if not trained, after accompanying boys nesting once or twice they will enter into the search with the greatest eagerness. Labourers occasionally make caps of dog skin, preserved with the hair on. Cats not uncommonly put a paw into the gins set for rabbits or rats. The sharp teeth break the bone of the leg, but if the cat is found and let out she will often recover – running about on three legs till the injured forefoot drops off at the joint, when the stump heals up. Foxes are sometimes seen running on three legs and a stump, having met with a similar disaster. Cats contrive to climb some way up the perpendicular sides of wheat ricks after the mice.

The sparrows are the best of gleaners: they leave very little grain in the stubble. The women who go gleaning now make up their bundles in a clumsy way. Now, the old gleaners used to tie up their bundles in a clever manner, doubling the straw so that it bound itself and enabled them to carry a larger quantity. Even in so trifling a matter there are two ways of doing it, but the ancient tradition and workmanship is dying out. The sheaves of corn, when set up in the field leaning against each other, bear a certain likeness to hands folded in prayer. By the side of cornfields the wild parsnip sometimes grows in great profusion. If dug up for curiosity the root has a strong odour, like the cultivated vegetable, but is small and woody. Everyone who has gathered the beautiful scarlet poppies must have noticed the perfect Maltese cross formed inside the broad petals by the black markings.

Beetles fly in the evening with such carelessness as to strike against people – they come against the face with quite a smart blow. Miserable beetles may sometimes be seen eaten almost hollow within by innumerable parasites. The labourers call those hairy caterpillars which curl in a circle

'Devil's rings' – a remnant of the old superstition that attributed everything that looked strange to demoniacal agency.

There is a tendency to variation even in the common buttercup. Not long since I saw one with a double flower; the petals of each were complete and distinct, the two flowers being set back to back on the top of the stalk. The stem of one of the bryonies withers up so completely that the shrinkage, aided by a little wind, snaps it. Then a bunch of red berries may be seen hanging from the lower boughs of a tree – a part of the stem, twined round, remaining there – the berries look as if belonging to the tree itself, the other part of the stem having fallen to the ground.

In clay soils the ivy does not attain any large size; but where there is some mixture of loam, or sand, it flourishes; I have seen ivy whose main stem growing up the side of an oak five inches in diameter, and had some pretensions to be called timber. The bulrush, which is usually associated with water, does not grow in a great many brooks and ponds; in some districts it is even rare, and it requires a considerable search to find a group of these handsome rushes. Water lilies are equally absent from certain districts. Elms do not seem to flourish near water; they do not reach any size, and a white, unhealthy-looking sap exudes from the trunk. Water seems, too, to check the growth of ash after it has reached a moderate size. Does the May bloom, which is almost proverbial for its sweetness, occasionally turn sour, as it were, before a thunderstorm? Bushes covered with this flower certainly emit an unpleasant smell sometimes quite distinct from the usual odour of the May.

The hedge is so intensely English and so mixed up in all popular ideas that it is no wonder it forms the basis of many proverbs and sayings – such as, 'The sun does not shine on both sides of the hedge at once', 'rough as a hedge', the verb 'to hedge', and so on. Has any attempt ever been made to cultivate the earthnut, pignut, or groundnut, as it is variously called, which the ploughboys search for and dig up with their clasp-knives? It is found by the small slender stalk it sends up, and insignificant white flower, and lies a few inches below the surface: the ploughboys think much of it, and it seems just possible that cultivation might improve it.

Rare birds do not afford much information as a rule – seen for a short time only, it is difficult to discover much about them. I followed one of the rarer woodpeckers one morning for a long time, but notwithstanding all my care and trouble could not learn much of its ways.

Even among cows there are some rudiments of government. Those who tend them say that each cow in a herd has her master (or rather mistress), whom she is obliged to yield precedence to, as in passing through a gateway. If she shows any symptoms of rebellion the other attacks her with her horns until she flies. A strange cow turned in among a herd is at once attacked and beaten till she gets her proper place – finds her level – when she is left in peace. The two cows, however, when they have ascertained which is strongest, become good friends, and frequently lick each other with their rough tongues, which seems to give them much satisfaction.

Dogs running carelessly along beside the road frequently go sideways: one shoulder somewhat in front of the other, which gives the animal the appearance of being ever on the point of altering his course. The longer axis of the body is not parallel to the course he is following. Is this adopted for ease? Because, the moment the dog hears his master whistle, and rushes forward hastily, the sidelong attitude disappears.

EIGHTEEN

Snake Lore

THERE ARE THREE KINDS OF SNAKES, according to the cottage people – water snakes, grass snakes, and black snakes. The first frequent the brooks, ponds, and withy beds; the second live in the mounds and hedges, and go out into the grass to find their prey; the third are so distinguished because of a darker colour. The cottage people should know, as they see so many during the summer; but they have simply given the same snake a different name because they notice it in different places. The common snake is, in fact, partial to the water, and takes to it readily. It does, however, seem to be correct that some individuals are of a blacker hue than the rest, and so have been supposed to constitute a distinct kind.

These creatures, like every other, have their favourite localities; and, while you may search whole fields in vain, along one single dry sandy bank you may sometimes find half a dozen, and they haunt the same spot year after year. So soon as the violets push up and open their sweet-scented flowers under the first warm gleams of the spring sunshine, the snake ventures forth from his hole to bask on the south side of the bank. In looking for violets it is not unusual to hear a rustling of the dead leaves that still strew the ground, and to see the pointed tail of a snake being dragged after him under cover.

In February there are sometimes a few days of warm weather (about the last week), and a solitary snake may perhaps chance to crawl forth; but they are not generally visible till later, and, if it be a cold spring, remain torpid till the wind changes. When the hedges have grown green, and the sun, rising higher in the sky, raises the temperature, even though clouds be passing over, the snakes appear regularly, but even then not till the sun has been up some hours. Later on they may occasionally be found coiled up in a circle two together on the bank.

In the summer some of them appear of great thickness – almost as big round as the wrist. These are the females, and are about to deposit their eggs. They may usually be noticed close to cowyards. The cattle in summer graze in the fields and the sheds are empty; but there are large manure heaps overgrown with weeds, and in these the snakes' eggs are left. Rabbits are fond of visiting these cowyards – many of which are at a distance from the farmstead – and sometimes bring forth a litter there.

When the mowers have laid the tall grass in swathes snakes are often found on them or under them by the haymakers, whose prongs or forks throw the grass about to expose a large surface to the sun. The haymakers kill them without mercy, and numbers thus meet with their fate. They vary very much in size – from eighteen inches to three feet in length. I have seen specimens which could not have been less than four feet long, and as thick as a rake handle. That would be an exceptional case, but not altogether rare. The labourers will tell you of much larger snakes, but I never saw one.

There is no subject upon which they make such extraordinary statements, evidently believing what they say, as about snakes. A man told me once that he had been pursued by a snake, which rushed after him at such a speed that he could barely escape; the snake not only glided but actually leaped over the ground. Now this must have been pure imagination: he fancied he saw an adder, and fled, and in his terror thought himself pursued. They constantly state that they have seen adders; but I am confident that no viper exists in this district, nor for some miles round. That they do elsewhere of course is well known, but not here; neither is the slow-worm ever seen.

The belief that snakes can jump – or coil themselves up and spring – is, however, very prevalent. They all tell you that a snake can leap across a ditch. This is not true. A snake, if alarmed, will make for the hedge; and he glides much faster than would be supposed. On reaching the shore or edge of the ditch he projects his head over it, and some six or eight inches of the neck, while the rest of the body slides down the slope. If it happens to be a steep-sided ditch he often loses his balance and rolls to the bottom; and that is what has been mistaken for leaping. As he rises up the mound he follows a zigzag course, and presently enters some small hole or a cavity in a decaying stole. After creeping in some distance he often meets with an obstruction, and has to remain half in and half out till he can force his way. He usually takes possession of a mouse hole, and does not seem to be able to enlarge it for additional convenience. If you put your stick on his head as he slips through the grass his body rolls and twists, and almost ties itself in a knot.

I have never been able to find a snake in the actual process of divesting his body of the old skin, but have several times disturbed them from a bunch of grass and found the slough in it. There was an old wall, very low and somewhat ruinous, much overgrown with barley-like grasses, where I found a slough several times in succession, as if it had been a favourite resort for the purpose. The slough is a pale colour – there is no trace on it of the snake's natural hue, and it has when fresh an appearance as if varnished – meaning not the brown colour of varnish, but the smoothness. A thin transparent film represents the eyes, so that the country folk say the snake skins his own eyes.

A forked stick is the best thing to catch a snake with: the fork pins the head to the ground without doing any injury. If held up by the tail – that is the way the country lads carry them – the snake will not let its head hang down, but holds it up as far as possible: he does not, however, seem able to crawl up himself; he is helpless in that position. If he is allowed to touch the arm he immediately coils round it. A snake is sometimes found on the roofs of cottages. The roof in such cases is low, and connected by a mass of ivy with the ground, overgrown too with moss and weeds.

The mowers, who sleep a good deal under the hedges, have a tradition that a snake will sometimes crawl down a man's throat if he sleeps on the ground with his mouth open. There is also a superstition among the haymakers of snakes having been bred in the stomachs of human beings, from drinking out of ponds or streams frequented by water snakes. Such snakes – green, and in every respect like the field snake – have, according to them, been vomited by the unfortunate persons afflicted with this strange calamity. It is curious to note in connection with this superstition the ignorance of the real habits of these creatures exhibited by people whose whole lives are spent in the fields and by the hedges.

Now and then a peculiar squealing sound may be heard proceeding from the grass, and on looking about it is found to be made by a frog in mortal terror: a snake has seized one of his hind legs and has already swallowed a large part of it. The frog struggles and squeals, but it is in vain. The snake, if once he takes hold, will gradually get him down. I have several times released frogs from this horrible position; they hop off apparently unhurt if only the leg has been swallowed. But on one occasion I found a frog half gone down the throat of its dread persecutor: I compelled the snake to disgorge it, but the frog died soon afterwards. The frog being a broad creature, wide across the back – at least twice the width of the snake – it appears surprising how the snake can absorb so large a thing.

In the nesting season snakes are the terror of those birds that build in low bushes. I have never seen a snake in a tree (though I have heard of their getting up trees), but I have seen them in hawthorn bushes several feet from the ground, and apparently proceeding along the boughs with ease. I once found one in a bird's nest: the nest was empty – the snake had doubtless had a feast, and was enjoying deglutition. In some places where snakes are numerous, boys when bird's-nesting always give the nest a gentle thrust with a stick first before putting the hand in, lest they should grasp a snake instead of eggs. The snake is also accused of breaking and sucking eggs – some say it is the hard-set eggs he prefers; whether that be so or no, eggs are certainly often found broken and the yolk gone. When the young fledglings fall out of the nest on to the ground they run great risks from snakes.

When sitting in a punt in summer, moored a hundred yards or more from shore, I have often watched a snake swim across the lake, in a place about 300 yards wide. In the distance all that is visible is a small black spot moving steadily over the water. This is the snake's head, which he holds above the surface, and which vibrates a little from side to side with the exertions of the muscular body. As he comes nearer a slight swell undulates on each side, marking his progress. Snakes never seem to venture so far from shore except when it is perfectly calm. The movement of the body is exactly the same as on land – the snake glides over the surface, the bends of its body seeming to act like a screw. They go at a good pace, and with the greatest apparent ease. In walking beside the meadow brooks, not everywhere, but in localities where these reptiles are common, every now and then you may see a snake strike off from the shore and swim across, twining in and out the stems of the green flags till he reaches the aquatic grass on the mud and disappears among it.

One warm summer's day I sat down on the sward under an oak, and leaned my gun against it, intending to watch the movements of a pair of woodpeckers who had young close by. But the drowsy warmth induced slumber, and on waking – probably after the lapse of some time – I found a snake coiled on the grass under one of my legs. I kept perfectly still, being curious to see what the snake would do. He watched me with his keen eyes as closely as I watched him. So long as there was absolute stillness he remained; the moment I moved, out shot his forked black tongue, and away he went into the ditch as rapidly as possible.

Some country people say they can ascertain if a hedge is frequented by snakes, by a peculiar smell: it is certain that if one is killed, especially if worried by a dog, there is an unpleasant odour. That they lie torpid during the winter is generally understood; but though I have kept an eye on the grubbing of many hedges for the purpose of observing what was found, I never saw a snake disturbed from his winter sleep. But that may be accounted for by their taking alarm at the jar and vibration of the earth under the strokes of the axe at the tough roots of thorn stoles and ash, and so getting away. Besides which it is likely enough that these particular hedges may not have been favourite localities with them. They are said to eat mice, and to enter dairies sometimes for the milk spilt on the flagstones of the floor.* They

---

*An extraordinary instance of this has been very kindly communicated to me by the writer of the following letter:

Dear Sir – *Apropos* of your reference to the notion that snakes drink milk, I think it may interest you to hear of a curious instance of this which occurred near here about three months ago. At Kingswood, the home farm of Kempstone (Mr J.H. Calcraft's place, near Corfe Castle), the dairyman noticed that something seemed to enter the dairy through a hole in the wall and take the milk. Thinking it was a mouse or rat, he set a common gin at the hole, and caught a snake every day until he had caught seventeen! Mr Calcraft would corroborate this. My informant was Mr Bankes, rector of Corfe Castle, who heard it from the dairyman himself.

Faithfully yours,

S.C. Spencer Smith.
Kingston Vicarage, Wareham, Dorset, October 27, 1878.

may often be found in the furrows in the meadows, which act as surface drains and are damp.

Frogs have some power of climbing. I have found them on the roofs of outhouses which were covered with ivy; they must have got up the ivy. Their toes are, indeed, to a certain degree prehensile, and they can cling with them. They sometimes make a low sound while in the ivy on such roofs; to my ear it sounds like a hoarse 'coo'. Cats occasionally catch frogs by the legs, and torment them, letting the creature go only to seize it again, and finally devouring it. The wretched creature squeals with pain and terror exactly as when caught by a snake.

No surer sign of coming rain than the appearance of the toad on the garden paths is known. Many cottage folk will tell you that the hundreds and hundreds of tiny frogs which may sometimes be seen quite covering the ground fall from the sky, notwithstanding the fact that they do not appear during the rain, but a short time afterwards. And there are certain places where such crowds of these creatures may be oftener found than elsewhere. I knew one such place; it was a gateway where the clay soil for some way round the approach had been trampled firm by the horses and cattle. This gateway was close to a slowly running brook, so slow as to be all but stagnant. Here I have seen legions of them on several occasions, all crowding on the ground worn bare of grass, as if they preferred that to the herbage.

Newts seem to prefer stagnant or nearly stagnant ponds, and are rarely seen in running water. Claypits from whence clay has been dug for brickmaking, and which are now full of water, are often frequented by them, as also by frogs in almost innumerable numbers in spring, when their croaking can be heard fifty yards away when it is still.

Labourers say that sometimes in grubbing out the butt of an old tree – previously sawn down – they have found a toad in a cavity of the solid wood, and look upon it as a great wonder. But such old trees are often hollow at the bottom, and the hollows communicate with the ditch, so that the toad probably had no difficulty of access. The belief in the venom of the toad is still current, and some will tell you that they have had sore places on their hands from having accidentally touched one.

They say, too, that an irritated snake, if it cannot escape, will strike at the hand and bite, though harmless. Snakes will, indeed, twist round a threatening stick; and, as it is evidently a motion induced by anger, the question arises whether they have some power of constriction. If so it is slight. In summer a few snakes may always be found by the stream that runs through the fields near Wick Farm.

This brook, like many others, in its downward course is checked at irregular intervals by hatches, built for the purpose of forcing water out into the meadows, or up to ponds at some distance from the stream at which the cattle in the sheds drink. Sometimes the water is thus led up to a farmstead; sometimes the farmstead is on the very banks of the brook, and the hatch is within a few yards. Besides the movable hatches, the stream in many places is crossed by bays (formed of piles and clay), which either irrigate adjacent meads or keep the water in ponds at a convenient level.

A lonely moss-grown hatch, which stands in a quiet shady corner not far from the lake, is a favourite resort of the kingfishers. Though these brilliantly coloured birds may often be seen skimming across the surface of the mere, they seem to obtain more food from the brooks and ponds than from the broader expanse of water above. In the brooks they find overhanging branches upon which to perch and watch for their prey, and without which they can do nothing. In the lake the only places where such boughs can be found are the shallow stretches where the bottom is entirely mud, and where the water is almost hidden by weeds. Willows grow there in great quantities, and some of their branches may be available; but then the water is hidden by weeds, and, being muddy at bottom, is not frequented by those shoals of roach the kingfisher delights to watch. So that the best places to look for this bird are on the streams which feed the mere (especially just where they enter it, for there the fish often assemble), and the streams that issue forth, not far from the main water.

This old hatch – it is so old and rotten that it is a little dangerous to cross it – is in the latter position, on the effluent, and is almost hidden among trees and bushes. Several hedges there meet, and form a small cover, in the midst of which flows the dark brook; but do not go near carelessly, for the bank is undermined by the water itself and by the water rats, while the real edge is concealed by long coarse grasses. These water rats are forever endangering the bay: they bore their holes at the side through the bank from above and emerge below the hatch. Out of one such hole the water is now rushing, and if it is not soon stopped will wear away the soil and escape in such quantities as to lower the level behind the hatch. These little beaver-like creatures are not, therefore, welcome near hatches and dams.

If you approach the cover quietly and step over the decayed pole that has been placed to close a gap, by carefully parting the bushes the kingfisher may be seen in his favourite position. The old pole must not be pressed in getting over it, or the willow 'bonds' or withes with which it is fastened to a tree each side of the gap will creak, and the pole itself may crack, and so alarm the bird. The kingfisher perches on the narrow rail that crosses the hatch about two feet above the water.

Another perch to which he removes now and then is formed by a branch, dead and leafless, which projects across a corner of the bubbling pool below. He prefers a rail or a dead branch, because it gives him a clearer view and better facilities for diving and snatching up his prey as it swims underneath him. His azure back and wings and ruddy breast are not equalled in beauty of colour by any bird native to this country. The long pointed beak looks half as long as the whole bird: his shape is somewhat wedge-like, enlarging gradually from the point of the beak backwards. The cock bird has the brightest tints.

In this pool scooped out by the falling water swim roach, perch, and sticklebacks, and sometimes a jack; but the jack usually abides near the edge out of the swirl. Roach are here the kingfisher's most common prey. He chooses those about four inches long by preference, and 'daps' on them the moment they come near enough to the surface. But he will occasionally land a much larger fish, perhaps almost twice the size, and will carry it to some distance, being remarkably powerful on the wing for so small a bird. The fish is held across the beak, but in flying it sometimes seems to be held almost vertically; and if that is really the case, and not an illusion caused by

the swiftness of the flight, the bird must carry its head a little on one side. If he is only fishing for his own eating, he does not carry his prey farther than a clear place on the bank. A terrace made by the runs of the water rat is a common table for him, or the path leading to the hatch where it is worn smooth and bare by footsteps. But he prefers to devour his fish either close to the water or in a somewhat open place, and not too near bushes; because while thus on the ground he is not safe. When feeding his young he will carry a fish apparently as long as himself a considerable distance.

One summer I went several days in succession to a hedge two fields distant from the nearest brook, and hid on the mound with a gun. I had not been there long before a kingfisher flew past, keeping just clear of the hedge, but low down and close under the boughs of the trees, and going in a direction which would not lead to a brook or pond. This seemed curious; but presently he came back again, uttering the long whistle which is his peculiar note. About an hour, perhaps less, elapsed when he returned again, this time carrying something in his beak that gleamed white and silvery in the sun – a fish. The next day it was the same, and the next. The kingfisher, or rather two of them, went continually to and fro, and it was astonishing what a number of fish they took. Never more than an hour, often less, elapsed without one or other going by. The fish varied much in size, sometimes being very small.

They had a nest, of course, somewhere; but, being under the idea that they always built near brooks or in the high banks often seen at the back of ponds, it was difficult for me to imagine where the nest could be. To all appearance they flew straight through a small opening in another hedge, at the corner of the two in fact, about two hundred yards distant. Presently it occurred to me that this might be an illusion, that the birds did not really pass through the hedge, but had a nest somewhere in that corner.

Just in the very angle was an old disused sawpit, formed by enlarging the ditch, and made some years before for the temporary convenience of sawing up a few heavy 'sticks' of timber that were thrown thereabouts. The sawpit, to prevent accidents to cattle, was roughly covered over with slabs of wood, which practically roofed it in, and of course darkened the interior. It was in this sawpit that the kingfishers had their nest, in what appeared to be a hole partly excavated by a rabbit. The distance from the hatch and brook was about four hundred yards, so that the parent birds had to carry the fish they captured nearly a quarter of a mile. The sawpit, too, was close to a lane used a good deal, though sheltered by a thick hedge from the observation of those who passed.

In another case I knew of, the kingfishers built in a mound overhanging a small stagnant and muddy pond, in which there were no fish, and which was within twenty paces of a farmhouse. The house was on a hill about three hundred yards from the nearest running stream. This little pond was full in wet weather only, and was constantly used by the horses, the cattle in the field that came almost up to the door, and by the tame ducks. Beside the pond was a woodpile, and persons were constantly passing it to and fro. Yet the kingfishers built there and reared their young; and this not only for one season, but for several years in succession. They had to bring all the fish they captured up from the brook, over the garden, and to pass close to the house. Why they should choose such a place is not easily explained, seeing that so many apparently more suitable localities were open to them along the course of the stream.

One summer I found a family of four young kingfishers perched in a row on a dead branch crossing a brook which ran for some distance beside a double-mound hedge. There was a hatch just there too, forcing the water into two ponds, one each side of the mound. The brook had worn itself a deep channel, and so required a hatch to bring it up to a level convenient for cattle. I had known for some time that there was a nest in that mound from the continued presence of the two old birds, but could not find it. But when the young could fly a little they appeared on this branch projecting almost over the falling water, and there they took up their station day after day. Every now and then the parents came with small fish, which they caught farther down the brook, for just in that place there were only a few perch and perhaps a tench or two. The colours are much less brilliant on the young birds, and they do not obtain the deep rich hues of their parents until the following spring. I have shot many young birds in the winter; they are by that time much improved in colour, but may be distinguished without difficulty from the full-grown bird.

Though so swift, the kingfisher is comparatively easy to shoot, because he flies as straight as an arrow; and if you can get clear of bushes or willow pollards he may be dropped without trouble. When disturbed the kingfisher almost invariably flies off in one favourite direction; and this habit has often proved fatal to him, because the sportsman knows exactly which way to look, and carries his gun prepared. Wherever the kingfisher's haunt may be, he will be found upon observation to leave it nearly always in the same direction day after day. He is, indeed, a bird with fixed habits, though apparently wandering aimlessly along the streams. I soon found it possible to predict beforehand in which haunt a kingfisher would be discovered at any time.

By noting the places frequented by these birds you know where the shoals of small fish lie, and may supply yourself with bait for larger fish. Often one of those great hawthorn bushes that hang over a brook is a favourite spot. The roots of trees and bushes loosen the soil, and deeper holes are often found under them than elsewhere, to which the fish resort. These hawthorn bushes, though thick and impenetrable above, are more open below just over the water; and there the kingfisher perches, and has also the advantage of being completely hidden from observation: if he only remained still in such places he would escape notice altogether. When passing such a bush on the *qui vive* for snipe, how many times have I seen a brilliant streak of azure shoot out from the lower branches and watched a kingfisher skim across the meadow, rising with a piping whistle over the distant hedge! Near millponds is a favourite place with these birds.

To that hatch which stands on the effluent brook not far from the mere, a coot or two comes now and then at night or in the early morning. These birds, being accused of devouring the young fry, are killed whenever they are met, and their eggs taken in order to prevent their increase; that is, of course, where the water is carefully preserved. Here they are not so persistently hunted. I have seen coots, and moorhens too, venture some distance up the dark arch of a culvert. Moorhens are fond of bridges and frequently feed under them. When alarmed, after diving, the moorhen does not always come right up to the surface, but merely protrudes its head to breathe.

One day I startled a moorhen in a shallow pond; instantly the bird dived, and I watched to see where it would come up, knowing that the moorhen cannot stay long under water, while there chanced to be scarcely any bushes or cover round the edge. After waiting some time, and wondering what had become of the bird, I fancied I saw some duckweed slightly agitated. Looking more carefully, it seemed as if there was something very small moving now and then just there – the spot was not more than fifteen yards distant. It was as if the beak of a bird, the body and most of the head quite hidden and under water, were picking or feeding among the duckweed. This continued for some few minutes, when I shot at the spot, and immediately a moorhen rose to the surface. As the pond was very shallow the bird must have stood on the bottom, and so resumed its feeding with the beak just above the surface.

NINETEEN

Course of the Brook

A PLACE WHERE THE BANK of the brook has been dug away so as to form a sloping approach to the water, in order that cattle may drink without difficulty, is much visited by birds in summer. Some cartloads of small stones originally thrown down to make a firm floor to the drinking place have in process of time become worn into sand, which the rain has washed into the water. This has helped to form a more than usually sandy bottom to the water just there. Then a bank of mud, or little eyot in the centre of the stream, thickly overgrown with flags, divides the current in two, and the swiftest section passes by the drinking place and brings with it more sand washed out from the mud; so that just at the edge there is a floor of fine sand covered with water, which six inches from shore is hardly an inch deep. This is just the bathing place in which birds delight, and here they come, accordingly, all the summer through, day after day.

Sparrows, starlings, finches (including the beautiful goldfinches), blackbirds, and so on, are constantly to and fro. Often several different species are bathing together. The wagtails, of course, are there. The wagtail wades into the water and stands there. Sometimes he has the appearance of scraping the bottom with his feet, as if to find food. Blackbirds are especially fond of this spot, and may be seen coming to it from the adjacent hedges. They like water, and frequently feed near it; a blackbird may often be found under the great hawthorn bushes which overhang the stream. Hawks may be seen occasionally following the course of the brook or perched on the trees that grow near; they are doubtless aware of the partiality for water shown by so many birds.

The fish have their own favourite places as the birds in the hedge, and after leaving the hatch there are none for some distance. Then the brook suddenly curves and forms a loop, returning almost upon itself something like the letter Ω. The tongue of land thus enclosed is broad at the top, and but two or three yards across at the bottom. There the current on either side is forever endeavouring to eat away the narrow neck, and forms two deep pools. Some few piles have been driven in on one side to check the process of disintegration, and a willow tree overhangs the pool there. By lying on the grass and quietly looking over the brink, the roach may be seen swimming in the deeper part, and where it shallows upstream is a perch waiting for what may come down. Where the water runs slowly on account of a little bay, there, in semi-darkness under the banks on the mud, are a few tench.

There are several jack pike not far off; but though they prey on the roach it is noticeable that, unless driven by someone passing by, they rarely go into these deep holes. The jack lies in shallower water and keeps close to the shore under shelter of the flags, or concealed behind the weeds. It is as if he understood that every now and then the shoal of roach will pass round the curve – going from one pool to the other – when they have to swim through the shallower water. Sometimes a solitary fish will shift quarters like this, and must go by the jack lying in ambush.

At the top of the tongue of land (which is planted with withy) another brook joins the first: this brook is very deep, and all but stagnant. In the quiet backwater here – close to and yet out of the swifter stream – is another haunt of the jack.

If alarmed, he does not swim straight up or down the centre of the current, but darts half a dozen yards in a slanting direction across the stream and hides under another floating weed. Then, if started afresh, he makes another zigzag, and conceals himself once more. At first he remains till you could touch him, if you tried, with a long stick; but at every remove he grows more suspicious, till at last as you approach he is off immediately.

Jacks lie a great deal in the still deep ponds that open off the brook or are connected with it by a deep ditch; they have been known to find their way up to a pond from the brook through a subterranean pipe which supplied it with water. Those that remain in the ponds are usually much larger than those found in the stream; these are often small – say, a pound to two

pounds in weight. In the spawning season, however, they come out from
the ponds and go up the brook in pairs or trios. They keep close together
side by side – the largest in the centre when there are three. The brook at
that time seems full of jacks; and to anyone who has been accustomed to
stroll along, it is surprising where they all come from.

Although the jacks lie in the quiet ponds most of the time, some of them
travel about a great deal, especially the smaller ones ranging from one to
two pounds. These will leap a bay or dam if it interrupts their voyaging
downstream. I have seen a young jack, about a foot long, leap over a bay,
and fall three or four feet on to the stony floor below, the stones scarcely
covered with water. The jack shot himself perhaps two feet and fell on his
side on the stones; there he lay quietly a minute or so, and then gave a
bound up, and, lighting in the current, went down with it. A small jack like
this will sometimes go out into the irrigated meadows, following the water
carriers for a long distance.

In quiet, sheltered places, where the water is clear but does not run too
swiftly, the 'minnie', as the stickleback is locally called, makes its nest
beside the bank. A small hole in the sand is excavated, and in this are laid
a number of tiny fibres such as are carried along by the stream, resembling
a miniature faggot. On these fibres the ova are deposited, and they are
then either purposely partly covered with sand by the minnie, or else the
particles that are brought down by the current gather over the bundle of
fibres and conceal it, excepting one small spot. There several of the slender
roots seem to slightly project, and they are kept clear of mud or sand so
as to answer the purpose of a doorway. I have watched these operations
many times, but never saw the minnie attempt to enter the nest; indeed, he
could not have done so, the opening not being large enough.

When the nest has reached this stage of completion it is easy to discover,
because the stickleback keeps watch before it, and at that season his breast
is of a bright crimson hue. He guards the nest with the greatest care, and
if he is tempted away for a minute by some morsel of food he is back
again immediately. If a tiny twig or fibre comes along and threatens to
catch against the nest, he removes it in his mouth, carrying it out into the

stream that it may be swept away. He also removes the sand whenever
it begins to accumulate overmuch. It would seem as if a current of fresh
water were essential to the ova, and that is why the opening of the nest
is so carefully kept from becoming choked up. After a while the fry come
forth – the most minute creatures imaginable, mere lines about half the
length of the fingernail. They play round the opening, and will retreat
within if alarmed.

Where the brook passes under a bridge of some size the current divides to
go through several small arches. There is here some fall, and the stream is
swift and bright, chafing round and bubbling over stones. Here the 'miller's
thumbs' are numerous – a bottom fish growing to about four inches in
length, and with a head enormously broad and large in proportion to its
body. They rarely rise from the mud or sand; they hide behind stones, their
heads buried in the sand, but their tails in sight. Every now and then they
change positions, swimming swiftly over the bottom to another spot. Their
voracity is very great, and they often disappoint the angler by taking his
bait. The cottage people are said to eat them.

The 'stwun loach' – stone loach, as the lads call it – hides also behind
and under stones, and may be caught by hand. These loach are apparently
capricious in their habits; certain spots abound with them, in others you
may search the stream in vain for a long distance. So, too, with the gudgeon:
I noticed in one brook I frequently passed that they never came up beyond
one particular bend, though there was no apparent difference in the soil or
in the stream itself. In the brook the jack do not seem to care much about
them; but in the lake above there are no gudgeon, and there a gudgeon is
a fatal bait. Nothing is so certain to take; the gudgeon will tempt the pike
there when an ordinary roach may be displayed before him without the
slightest effect.

A flood which brings down a large quantity of suspended mud and sand
discolouring the water attracts the fish: they are looking for food. But too
much mud compels them to shift their quarters. This is well known to those
who net the stream. They stretch the net across the brook a few yards below
a bridge or short culvert – places much haunted by fish. Then the bottom

of the stream above the culvert is thoroughly stirred up with a pole till the water is thick with mud, and this, passing through the culvert (where the pole cannot be used and the fish would otherwise be safe), forces them to descend the stream and enter the net. Probably they attempt to swim upstream first, but are deterred by the pole thrust under the water, and then go down. It is said that even eels, who like mud, will move if the volume of mud sent through is thick enough and continued sufficiently long.

The fact that a little stirring of the bottom attracts fish is made use of along the Thames to attract bait for those night-lines which are the detestation of the true angler. The bait catcher has a long pole, at the end of which are iron teeth like a rake. With this he rakes up the mud, waits a few seconds, and then casts a net, which generally brings some minnows or other small fish to shore. These fish are then placed in a bucket, and finally go on the night lines.

The ditches as they open on the brook are the favourite resorts of all aquatic life, and there most of the insects, beetles, etc., that live in the water may be discovered. They form, too, one of the last resorts of the reeds; these beautiful plants have been much diminished in quantity by the progress of agriculture. One or two great mounds by the brook can show a small bed still, and here and there a group grows at the mouth of these deep ditches, on the little delta formed of the sand, mud, and decaying twigs brought down. I have cut them fifteen feet in length. Some people, attracted by the beauty of the feathery heads of these reeds, come a considerable distance to get them. I have made pens of them: it is possible to write with such pens, and they are softer than quills, but on account of that softness quickly wear out.

A woodcock may occasionally be flushed from such a ditch in winter. Woodcocks are fond of those ditches down which there always trickles a tiny thread of water – hardly so much as would be understood by the term streamlet – coming from a little spring which even in severe frosts is never frozen. Even when the running brook is frozen such little springs are free of ice, and so, too, is the streamlet for some distance.

From the bed of the brook proper the reeds are gone – they have taken refuge in nooks and corners. This is probably accounted for by the periodical cleaning out of the brook – not annually, but every now and then, in order to prevent the flooding which would be caused by the accumulation of mud and sand. The roots of the flags seem to withstand this rude treatment; but many other water plants cannot, and are consequently only found in places which have not been disturbed for many years.

There is as much difference in ponds as in hedges, so far as inhabitants are concerned. Many fields and hedges seem comparatively deserted, while others are full of birds; and so of several ponds which do not apparently vary much – one is a favourite haunt of fish, and another has not got a single fish in it. One pond particularly used to attract my attention, because it seemed devoid of any kind of life: not even a stickleback could be found in it, though they will live in the smallest ditches, and this pond was fed by a brook in which there was fish. Not even a newt lived in it – it was a miniature Dead Sea. Another pond was remarkable for innumerable water snails. When the wind blew hard they sometimes lined the lee shore to which they had drifted.

The herons at the same time are the largest and most regular visitors to the mere out of which the brook flows. One or more may generally be found there at some time of the day all the year round; but there is a remarkable diminution in their numbers during the nesting season. The nearest herony must be about thirty miles distant, which probably explains their absence at that time. It also happens that just before the summer begins the mere is usually at its greatest height: the water is deep almost everywhere, and there are fewer places where the herons could fish with success.

They fly at a great height in the air, and a single stroke of the huge wings seems to propel the bird a long distance; so that though at first sight they appear to move very slowly, the eye being deceived by the slow stroke of the wings, they really go at a good pace. They do not seem to have any regular hours of visiting the lake – though more seem to arrive in the afternoon – but they have distinct lines of flight along which they may be expected to come. In winter, however, they show more regularity, going down from the lake to the water meadows in the evening, and returning

in the early morning – that is, supposing the lake to be open and free from ice. If the shores are frozen, a heron or two may be found in the water meadows all day.

In the autumn, after a dry summer, is the best time to watch them. The water is then low; numerous small islands appear, and long narrow sandbanks run out fifty or sixty yards, with shoals on either side. After a very dry season the level of the water is so much reduced that in the broadest (and shallowest) part the actual strand where the water begins is a hundred yards or more from the nearest hedge. This is just what the heron likes, because no one can approach him over that flat expanse of dried mud without being immediately detected. I have seen as many as eight herons standing together in a row on one such narrow sandbank in the daytime, in regular order like soldiers: there were six more on adjacent islands. They were not feeding – simply standing motionless. As soon as it grew dark they dispersed, and ventured then down the lake to those places near which footpaths passed.

But although the night seems the heron's principal feeding time, he frequently fishes in the day. Generally, his long neck enables him to see danger, but not always. Several times I have come right on a heron, when the banks of a brook were high and the bushes thick, before he has seen me, so as to be for the moment within five yards. His clumsy terror is quite ludicrous: try how he will he cannot fly fast at starting; he requires fifty yards to get properly under way.

What a contrast with the swift snipe, that darts off at thirty miles an hour from under your feet! The long hanging legs, the stretched-out neck, the wide wings and body, seem to offer a mark which no one could possibly miss: yet, with an ordinary gun and snipe-shot, I have had a heron get away safely like this more than once. You can hear the shot rattle up against him, and he utters a strange, harsh, screeching 'quaack', and works his wings in mortal fright, but presently gets halfway up to the clouds and sails away in calm security. His neck then seems to drop down in a bend, the head being brought back as he settles to his flight, so that the country people say the heron often carries a snake.

The mark he offers to shot is much less than would be supposed; he is all length and no breadth; the body is very much smaller than it looks. But if you can stalk him in the brook till within thirty or forty yards, and can draw a 'bead' on his head as he lifts it up every now and then to glance over the banks, then you have him easily; a very small knock in the head being sufficient to stop him.

The tenacity of life exhibited by the heron is something wonderful: though shot in the head, and hung up as dead, a heron will sometimes raise his neck several hours afterwards. To wring the neck is impossible – it is like leather or a strong spiral spring; you cannot break it, so that the only way to put the creature out of pain is to cut the artery; and even then there are signs of muscular contraction for some time. A labourer once asked me for a heron that I had shot; I gave it to him, and he cooked it. He said he boiled it eight hours, and that it was not so very fishy! But even he could not manage the neck part.

This bird must have a wonderful power of sight to catch its prey at night, and out of some depth of water. In severe winter weather, when the lake is frozen, herons evidently suffer much. Most of them leave, probably for the rivers which do not freeze till the last; but one or two linger about the water meadows till they seem to despair of catching anything, and will alight in the centre of a large pasture field where there is no water, and stand there for hours disconsolate. I suspect that the herons in wintertime that come to the ponds do so for the fish which lie at the bottom on the mud packed close together, that is, when the water is not deep. It is said that when ice protects the fish, herons eat the frogs in the water meadows; but they can scarcely find many, for though I have been over the water meadows day after day for snipe, I seldom saw a frog about them here.

When the level of the mere is very low, after a peculiarly dry season, it is a good time to observe the habits of many other creatures. There are always one or more crows about the neighbourhood of the lake; but at such times a dozen or so may be seen busily at work along the shore. They prey on the mussels, of which there are great numbers in the lake. Any one passing by the water when it is so shallow can hardly fail to notice long narrow grooves in the sand of the bottom. These grooves begin near the edge – perhaps within a foot of it – and then run out into the deeper part. By following these with the eye, the mussel may often be seen in a foot or two of water – sometimes open, but more generally closed. The groove in the sand is caused by the keel of the shell as the creature moves.

There are hundreds of these tracks; the majority appear to run from shallow to deep water, but there are others crossing and showing where the mussel has travelled. One may occasionally be seen in the act of moving itself, and making the groove in the sand. But they seem as a rule to move most at night, and to approach the shore closest in the darkness. In the deep water they are safe; but near the edge the crows pounce on them.

Besides those that are eaten on the shore, numbers of mussels are carried up on the rising ground where the turf is short and the earth hard. Until stepped on and broken, the two halves of the shell are usually complete, and generally still attached, showing that the crow has split the shell open skilfully. They range from two or three to nine inches in length. The largest are much less common; those of five or six inches are numerous. Some of the old-fashioned housewives use a nine-inch mussel shell, well-cleaned, as a ladle for their sugar jars.

Now and then, at long intervals, an exceptionally dry season so lowers the level of the mere that all the shallower parts become land, and are even passable on foot, though in places quicksands and deep fine mud must be carefully avoided. The fish that previously could enjoy a swim of some three-quarters of a mile are then forced to retire to one deep hole only a few acres in extent. Now commences a reign of terror, difficult to convey.

These waters have not been netted for years, and consequently both pike and perch have increased to an extraordinary degree, and many of them have attained huge proportions. Pike of six pounds are commonly caught; eight, ten, twelve, and fourteen-pound fish have often been landed. There was a tradition of a pike that weighed a quarter of a hundredweight, but one day the tradition was put into the shade by the capture of a pike that scaled a little over thirty pounds. There are supposed to be several more such monsters of the deep, since every now and then some labourer passing by on a sunny day, when jack approach the shore and bask near the surface, declares that he has seen one as big as a man's leg. But about the vast number of ordinary-sized jack there can be no doubt at all; since anyone may see them who will stroll by the water's edge on a bright, warm day, taking care to walk slowly and not to jar the ground or let his shadow fall on the water before he can glance round the willows and bushes. Jack may then be seen basking by the weeds.

When an exceptionally long continuance of dry weather forces all the fish to retire to the few acres of water that remain, then these voracious brutes do as they please with the other fish, and the roach especially suffer. Every two or three minutes the fry may be seen leaping into the air in the effort to escape, twenty or thirty at a time, and falling with a splash. The rush of hundreds and hundreds of roach causes a wave upon the surface which shows the course they take. This wave never ceases: as soon as it sinks here it rises yonder, and so on through the twenty-four hours, day and night.

The miserable fish, flying for their lives, speed towards the shallow water, and often, unable to stop themselves, are carried by their impetus out on the mud and lie there on the land for a few seconds till they leap back again. Even the jack will sometimes run himself aground in the eagerness of his pursuit. Looking over the pool, the splash of the falling fish as they descend after the leap into the air may be heard in several directions at once, and the glint of their silvery sides in the sunshine is at the same time visible. At night it is clear the same thing is going forward, for the splashing continues, though the wave raised by the panic-stricken crowds cannot be distinguished in the darkness.

It is curious to notice how the solitary disposition of the jack shows itself almost as soon as he comes to life. While the fry of most other fish swim in shoals, sometimes in countless numbers, the tiny jack, hardly so long as one's little finger, lurks all alone behind a stone which forms a miniature harbour. On a warm day almost every such place has it youthful pirate. Notwithstanding the terror of the roach when pursued, they will play about apparently without the slightest fear when the pike is basking in the sun with his back all but on a level with the surface – that is, when the lake is at its ordinary height. It is as if they knew their tyrant was enjoying his siesta.

These roach literally swarm. For a distance of some hundred and fifty yards the water for seven or eight feet from shore is simply a moving mass of roach. They crowd up against the stones, get underneath them and behind them, enter every little creek and interstice, and are so jammed by their own numbers that they may easily be caught by hand. In their anxiety to secure a place they crush against each other and splash up the water. This impulse only lasts a day or two in its full vigour, when the multitude gradually retires into deeper water.

When thus spawning the roach are preyed on by rats – not the water rat, but the house or drain rat. There are always a few of these about the lake, and they grow to an enormous size. They destroy the roach in great numbers. I have seen the sand strewn with dead fish opposite and leading up to their holes; for they catch and kill many more than they can eat, or even have time to carry away. I have shot at these great rascals when they have been swimming fifty yards from shore, and I strongly suspect them of visiting the nests of moorhens and other waterfowl with felonious purposes. They catch fish at any time they see a chance, but are most destructive during the spawning season, because then the roach come within reach.

Such rats, too, haunt the ditches and mounds, and are as dangerous to all kinds of game as any weasel, crow, or hawk.

Tench lie in the deep muddy holes. With the exception of the tench, the greater number of the fish in this mere haunt the sandy and stony shores. When the lake is full there are broad stretches of water which are shallow and where the bottom is mud. You may look here in vain for fish: of course there are some; but as you glide over noiselessly in a punt, gazing down into the water as you drift before the gentle summer breeze, you will not see any of those shoals that frequent the other shores where the bottom is clearer. Other favourite places are where the brooks run in and where there are sudden shallows in the midst of deep water. The contour and character of the bottom seem to affect the habits of fish to a large extent; consequently those who are aware of the form of the bottom are usually much more successful as fishermen.

TWENTY

Wildfowl of the Lake

THE 'SUMMER SNIPE', or sandpiper, comes to the lake regularly year after year, and remains during the warm months. About a dozen visit the shallow sandy reaches running along the edge of the water, when disturbed flying off just above the surface with a plaintive piping cry. They describe a semi-circle, and come back to the shore a hundred yards farther on; and will do this as many times as you like to put them up. Sometimes they feed in little parties of two or three: sometimes alone. No other place for some distance is visited by the sandpiper: none of the ponds or brooks; only the lake.

In summer but a few species of birds remain on this piece of water. Only two or three wild ducks stay to breed: their nests are not found on the mere itself, but in the ponds adjacent. One small pond fed by the lake and communicating with it – dug where the muddy shore would otherwise prevent cattle approaching the shallow water – a quiet spot almost surrounded by bushes, is a favourite nesting place. The brooks that run in are occasionally used by ducks in the same way, and one of the large ditches which is full of flags and rushes and well sheltered is now and then selected. But the ducks do not breed in any number, though they used to do so within living memory.

The coots cannot be overlooked in spring; they chase each other to and fro over the surface in the liveliest manner, and their nests are common. Moorhens, of course, are here in numbers. Why is it that they never seem to learn wisdom in placing their nests? Whether in the lake, in the ponds, or brooks they exhibit the same lack of foresight as to changes of level in the water; so that frequently their nests are quite drowned out. Occasionally in the brooks the nest is floated bodily down the stream by a sudden rise. These mishaps they might easily avoid by placing them a little higher up the bank.

In the lake there are several acres of withy bushes which when the water is low are on dry land, but in spring and early summer stand five or six feet deep. This is a favourite nesting place with the coots: and they show the same neglect of the teachings of experience; for their nests are placed almost on the water, and if it rises, as it often does, they are flooded.

It is said that otters used to come to the mere many years ago; but they have never done so lately, though stories of their having been seen are frequent. One summer the story was so positive and so often repeated that I made a thorough search, and found that it originated in the motions of a large diving bird. This bird swam under water with wonderful rapidity, and often close to the surface, so that it raised a wave and could be traced by it. This was the supposed otter. The bird was afterwards shot, but its exact species does not seem to have been satisfactorily ascertained. Several kinds of divers, however, have without doubt been killed. Grebes are often shot.

Occasionally seabirds come – particularly a species locally called the 'sea swallow', which frequently appears after rough winds and remains flying about over the water for a week or more. Six or eight of these are sometimes seen at once. The common gull comes at irregular intervals, generally in the winter or spring; it is said to foretell rough weather. Occasionally a gull will stay some time, and I have seen them also in the water meadows. Considering the distance from the sea, the gull cannot be called an uncommon bird here.

Towards winter the wild ducks return; and during all the cold months a flock of them, varying in number, remains. They are careful to swim during the day in the centre of the very widest part of the lake, far out of gunshot; at night they land to feed along the shore. Teal, and sometimes widgeon also, visit the place. Once now and then wildfowl come in countless numbers: it is said to be when they are driven south by severe weather. On one occasion I saw the lake literally black – they almost covered it for a length of half a mile and a breadth of about a quarter. It was a sight not to be quickly forgotten; and the noise of their wings as vast parties every now and then rose and wheeled around was something astonishing. They only stayed a few days.

How many times I have endeavoured to trace the V said to be formed by duck while flying, and failed to detect it! They fly, it is true, in some sort of order, but those that come to the mere here travel rather in a row or line, slanting forwards, something like what military men call in echelon. The teal seem much bolder than the wild duck: they are often shot as they rise out of the brooks; but the ducks very rarely go to the brooks at all, and can still more rarely be approached when they do. They swim in the water carriers in the great irrigated meadows, but are careful to remain far out of range; so that the only way to shoot them by day is for two or more sportsmen to post themselves behind the hedges in different places while a third drives them up.

The first snipes are seen generally in the arable lands, afterwards round the lake – the muddy shores by choice – and finally in the brooks. As the winter advances they seem to quit the lake in great part and go down to the brooks. A streamlet that runs through a peaty field is a favourite spot. The little jack snipe frequent the water carriers in the irrigated meadows and the wet furrows. When the lake is frozen over the wild duck stand on the ice in the daytime for hours together, leaving the marks of their feet on it.

In walking along the shore lines of drift may be noticed, marking the height to which the waves driven by the wind have carried the floating twigs, weeds, and leaves: just as along the sea the beach is formed into terraces by the changing height of the tides. The shallower parts of the lake are so thickly grown in summer with aquatic weeds that a boat can only be forced through them with the utmost difficulty. Some of these grow in as much as eight or even ten feet of water. On the shore, where it is marshy, the mare's tail flourishes over some acres: there is often a slight marshy odour here, which increases as the foot presses the yielding mud.

When the water is low in autumn these are mown, and, with the aquatic grasses at the edge and the rushes, made into the roughest kind of hay imaginable. The coarser parts are used as litter; the best is mixed with fodder and eaten by cattle. Many waggon loads are thus taken away, but as many more remain; and in walking over the spongy ground a smart 'pop' is continually heard: it is caused by the sudden compression of air under the foot in the mare's tails lying about; for their stems are hollow, and have knots at regular intervals.

After a continuance of the wind in one quarter for a few days – south or south-west – the opposite shores are lined with such weeds carried across, together with great quantities of dead branches fallen from the trees and willows. So that on a small scale the same thing happens as with the driftwood of the ocean; and, indeed, by studying the action of natural forces as exhibited in our own pools and brooks, it becomes much easier to comprehend the gigantic operations by which the surface of the earth is altered.

For instance, the north-eastern edge of the water is continually encroaching on the land; eating away the sandy soil, showing that the prevalent winds are south and west. The waves, thrown against the shore with the force they have acquired in rolling six or seven hundred yards, wash away the earth and undermine the bank, forming a miniature cliff or precipice, the face of which is always concave, projecting a little at the foot and also at the top. So much is this the case that an unwary person walking too near the edge may feel the sward suddenly yield and find it necessary to scramble off before a few hundredweights of earth subside into the water.

In this process the loamy part of the earth is dissipated, or rather held in suspension, while the small stones and ultimately the heavier sand fall to the bottom and form the sandy floor preferred by the fish. The loam discolours the water during a storm for several yards out to sea, so that in a boat passing by you know by the hue of the waves when you are approaching the dangers of the cliffs. This continuous eating away of the earth proceeds so fast that an old hollow oak tree now stands – at what may be called the high tide of summer – so far from the strand that a boat may pass between.

Like a wooded island the old oak rears itself up in the midst; the waves break against it, and when there is but a ripple the sunlight glancing on the water is reflected back, and plays upon the rugged trunk, illuminating it with a moving design as the wavelets roll in. The water is so shallow at the edge that the shadows of the ridges of the waves follow each other over the sandy floor. They reflect the bright rays upon the tree trunk, where they weave a beautiful lace-like pattern – beneath, their own shadows glide along the sand. That sand, too, is arranged by the ripple in slightly curved lines. These wave marks, though so slight that with the hand you may level fifty at a sweep, have yet sometimes proved durable enough to tell the student after many centuries where water once has been. Under the foundations of some of the oldest churches – the monks loved to build near water – the wave mark has been found on the original soil. In a hollow of the old oak starlings have made their nest and reared their young in safety for several seasons. They seem to understand that the water gives them protection, for their nest would not be out of reach were the tree on land.

Just as at the seashore the wave curls over in an arch as it comes in before dissolving in surf and spray; so here when a gale is blowing these lesser waves, as they reach the shelving strands also curl over. In the early morning, as the sun begins to acquire some strength, the white mists sweep over the surface and visibly melt and disappear. One hot summer, when the lake was full, and kept so artificially by the hatches and dams, I found by observation that its level sank nearly half an inch every day. This was the more striking because there was at the same time an influx more than enough to repair the loss from leakage. Now the evaporation of half an inch of water over such a width of surface meant the ascension into the atmosphere of many thousands of gallons; and thus even this insignificant pool might form a cloud of some magnitude in a few days. What immense vapours may then arise from the surface of the ocean!

Sometimes a winter's morning is, I think, almost as beautiful as summer, when the ice is thick with the sharp frost, and the sun shines in a blue sky free from clouds. One such morning, while putting on my skates, I happened to look up, and was surprised to see a bird of unusual appearance, and large size, soaring slowly overhead. I immediately recognised an eagle; and that was the solitary occasion on which I ever saw one here. The bird remained in sight some time, and finally left, going south-east towards the sun.

On the afternoon of the day before the beginning of the frost the wind gradually sinks, and the dead leaves which have been blown to and fro settle in corners and sheltered places. As the sun sets all is still, and there is a sense of freshness in the air. Then the logs of wood thrown on the fire burn bright and clear – the surface of a burning log breaks up into small irregular squares and the old folk shake their heads and say, 'It will freeze.' As the evening advances the hoofs of horses passing by on the road give out a sharp sound – a sign that the mud is rapidly hardening. The grass crunches underfoot, and in the morning the elms are white with rime; icicles hang from the thatch, and the ponds are frozen.

But there is nothing so uncertain as frost; it may thaw, and even rain, within a few hours, and, on the other hand, even after raining in the afternoon, it may clear up about midnight, and next morning the ice will be a quarter of an inch thick. Sometimes it will begin in so faint-hearted a fashion that the ground in the centre of the field is still soft, and will 'poach' under the hoofs of cattle, while by the hedge it is hard. But by slow degrees the cold increases, and ice begins to form. Again, it will freeze for a week and yet you will find very little ice, because all the while there has been a rough wind, and the waves on the lake cannot freeze while in motion. So that a long frost is extremely difficult to foresee.

But it comes at last. Two really sharp frosts will cause ice thick enough to bear a lad at the edge of the lake; three will bear a man a few yards out; four, and it is safe to cross; in a week the ice is between three and four inches thick, and would carry a waggon. The character of ice varies; if some sleet has been falling – or snow, which facilitates freezing – it is thick in colour; if the wind was still it is dark, sleek, perfectly transparent. It varies, however, in different places, in some having a faint yellowish hue. There are always several places where the ice does not freeze till the last – breathing-holes in which the ducks swim; and where a brook enters it is never quite safe.

The snipes come now to the brook and water meadows. Following the course of the stream, fieldfares and redwings rise in numbers from every hawthorn bush, where they have been feeding on the peggles. Blackbirds start out from under the bushes, where there is perhaps a little moist earth still. The foam where there is a slight fall is frozen, and the current runs under a roof of ice; the white bubbles travel along beneath it. The moorhens cannot get at the water; neither can the herons or kingfishers. The latter suffer greatly, and a fortnight of such severe weather is fatal to them.

I recollect walking by a brook like this, and seeing the blue plumage of a kingfisher perched on a bush, I swung my gun round ready to shoot as soon as he should fly, but the bird sat still and took no notice of my approach. Astonished at this – for the kingfisher sat in such a position as easily to see any oncoming, and these birds generally start immediately they perceive a person – I walked swiftly up opposite the bush. The bird remained on the bough. I put out the barrel of my gun and touched his ruddy breast with the muzzle; he fell on the ice below. He had been frozen on his perch during the night, and probably died more from starvation than from cold, since it was impossible for him to get at any fish.

More than once afterwards the same winter I found kingfishers dead on the ice under bushes, lying on their backs with their contracted claws uppermost, having fallen dead from roost. Possibly the one found on the branch may have been partly supported by some small twig.

That winter snow afterwards fell and became a few inches thick, drifting in places to several feet. Then it was the turn of the other birds and animals to feel the pain of starvation. In the meadows the tracks of rabbits crossed and recrossed till the idea of following their course had to be abandoned. At first sight it seemed as if the snow had suddenly revealed the presence of a legion of rabbits where previously no one had suspected the existence of more than a dozen. But in fact a couple of rabbits only will so run to and fro on the snow as to cover a meadow with the imprints of their feet – looking everywhere for a green blade.

Yet they only occasionally scratch away the snow and so get at the grass. Though the natural instinct of rabbits is to dig, and though here and there a place may be seen where they appear to have searched for a favourite morsel, they do not seem to acquire the sense of systematically clearing snow away. They then bark ash – and, indeed, nearly any young sapling or tree – and visit gardens in the night, as the hares do also. They creep about along the mounds, being driven by hunger to search for food all day instead of remaining part of the time in the burrows.

As to the hares, little more than a week of deep snow cripples their strength: they will run but twenty or thirty yards, and may be killed occasionally with a stick or captured alive. They are even more helpless than rabbits, because the latter still have holes to take refuge in from danger; but the hare while the snow lasts is a wretched creature, and knows not where to turn. Birds resort to the cattle sheds to roost; among them the blackbirds, who usually roost in the hedges. Birds come to the houses and gardens in numbers because the snow is there cleared away along the paths.

During severe weather the water meadows are the most frequented places. They are rarely altogether frozen. If in the early morning there are sheets of ice, by noonday a great part will be flooded an inch or two deep, the water rising over the ice, and forced by it to spread farther, softening the ground at the sides. Thrushes and blackbirds come to the hedges surrounding these meadows; the fieldfares and redwings are there by hundreds, and fly up to the trees if alarmed.

The old folks say that the irrigated meadows (and other open waters) do not freeze in the evening till the moon rises; a bright clear moon is credited with causing the water to 'catch' – that is, the slender, thread-like spicules form on the surface, and, joining together, finally cover it. It is, of course, because the water meadows are long before altogether frozen so that the duck and teal come down to them. When the brooks are frozen is almost the only time when the dabchick can be got to rise: at other times this bird will dive and redive, and double about in the water, and rather be caught by the spaniels than take wing. But when the ice prevents this they will fly. Wood pigeons go to the few places that remain moist, and also frequent the hawthorn bushes with the fieldfares. They seem fond of trees that are overgrown with ivy, probably for the berries.

The fish are supposed to go down upon the mud; but the jacks certainly do the reverse; they may be seen lying just beneath the ice, and apparently touching it with their backs. They seem partly torpid. In open winters, such as we have had of recent years, the hedge fruit remains comparatively untouched by birds: from which it would appear that it is not altogether a favourite food.

The country folk, who are much about at night and naturally pay great heed to the weather, are persuaded that on rainy nights more shooting stars are seen that when it is bright and clear. The kind of weather they mean is when scudding clouds with frequent breaks pass over, now obscuring and now leaving part of the sky visible, and with occasional showers. These shooting stars, they say, are just above the clouds, and are mere streaks of light: by which they mean to convey that they have no apparent nucleus and are different from the great meteors which are sometimes seen.

I have myself been often much interested in the remarkable difference of the degree of darkness when there has been no moon. There are nights when, although the sky be clear of visible cloud and the stars are shining, it is, in familiar phrase, 'as black as pitch'. The sky itself is black between the stars, and they do not seem to give the slightest illumination. On the other hand, there are nights without a moon when it is (though wintertime) quite light. Hedges and trees are plainly visible; the road is light, and anything approaching can be seen at some distance, and this occasionally happens though the sky be partly clouded. So that the character of the night does not seem to depend entirely upon the moon or stars. The shepherds on the hills say that now and then there comes an intense blackness at night which frightens the sheep and makes them leap the hurdles.

When logs of timber are split for firewood they are commonly stacked 'four square', and occasionally such a stack, four or five feet high, may be seen all aglow with phosphorescence. Each individual split piece of wood is distinctly visible – a pale faintly yellow light seems to be emitted from its surface. At the same time the ends of the faggot sticks projecting from the adjacent stack of faggots also glow as if touched with fire. So vivid is the light that at the first glance it is quite startling – as if the whole collection of wood were just on the point of bursting into flame. In passing old hollow trees sometimes they appear illuminated from within: the light proceeds from the decaying touchwood. Old willow trees are sometimes streaked with such light from the top to the foot of the trunk. As this phosphorescence is only occasional, it would seem to depend on the condition of the atmosphere.

I once noticed what looked like a glow-worm on a window at night, but there was no glow-worm there; the light was of a pale greenish hue. In the morning an examination showed that the linen was decayed and almost rotten just in that particular spot, and it had slightly turned colour.

The *ignis fatuus* is almost extinct; so much so that Jack-o'-the-Lantern has died out of the village folklore. On one occasion, however, I saw what at a distance seemed a bright light shining in a ditch where two hedges met. Thinking some mischief was going on, I went to the spot, when the light disappeared; but on retiring, after a search which proved that no one was about, it came into view again. A second time I approached, and a second time the light died out. A few nights afterwards it was there again, and must clearly have been some kind of *ignis fatuus*. There was a small quantity of stagnant water in the ditch, and a good deal of rotten wood – branches fallen from trees.

One of the most interesting phenomena in connection with the weather seems to me to be the radiation of clouds. It appears to be more commonly visible in the evening, and, when fully developed, there is a low bank on the horizon, roughly arched, from which streamers of cloud trail right across the sky, through the zenith and down to the horizon opposite. Near each horizon these streamers or lines almost touch; overhead they are wider apart – an effect of perspective, I suppose. Often the lines do not stretch so far, hardly to the zenith, where they spread out like a fan. If the sun has gone down, and the cloud chances to be white, these lines greatly resemble the aurora borealis, which takes the same form, and, when pale, can scarcely be distinguished from them, except for the streamers shooting – now extending, now withdrawing – while the cloud streamers only drift slowly. Sometimes there is but one line of cloud, a single streamer stretching right across the sky. So far as I have been able to observe, this radiation is usually followed by wind blowing in a direction parallel to the course of the streamers.

Once while walking in winter I was overtaken by a storm of rain, and took shelter behind a tree, which for some time kept me perfectly dry.

But suddenly there came an increase of darkness, and, glancing round, I saw a black cloud advancing in the teeth of the wind, and close to the earth. The trees it passed were instantly blotted out, and as it approached I could see that in the centre it bulged and hung down – or rather slightly slanted forward – in the shape of an inverted cone with the apex cut off. This bulging part was of a slaty black, and the end travelled over the earth not higher than half the elevation of an ordinary elm. It came up with great speed, and in a moment I was completely drenched, and the field was flooded. It did not seem so much to rain as to descend in a solid sheet of water; this lasted a very short time, and immediately afterwards the storm began to clear. Though not a perfect water spout, it was something very near it. The tree behind which I had taken shelter stood near a large pond, or mere; and I thought at the time that that might have attracted the cloud. The field quite ran with water, as if suddenly irrigated, but the space thus flooded was of small area – about an acre.

The haymakers sometimes talk of mysterious noises heard in the very finest weather, when it is still and calm, resembling extremely distant thunder. They were convinced it was something 'in the air'; but I feel certain it was the guns of the fleet exercising at sea. In that case the sound of the explosion must have travelled over fifty miles in a direct line – supposing it to come from the neighbourhood of the nearest naval station. I have found by observation that thunder cannot be heard nearly so far as the sound of cannon. I doubt whether it is often heard more than ten miles. Some of the old cottage folk are still positive that it is not the lightning but the thunder that splits the trees; they ask if a great noise does not make the windows rattle, and want to know whether a still greater one may not rive an oak. They allow, however, that the mischief is sometimes done by a thunderbolt.

Please contact Little Toller Books
to join our mailing list or for more information
on current and forthcoming titles.

Nature Classics Library

THE JOURNAL OF A DISAPPOINTED MAN *W.N.P. Barbellion*
THROUGH THE WOODS *H.E. Bates*
MEN AND THE FIELDS *Adrian Bell*
ISLAND YEARS, ISLAND FARM *Frank Fraser Darling*
A SHEPHERD'S LIFE *W.H. Hudson*
WILD LIFE IN A SOUTHERN COUNTY *Richard Jefferies*
FOUR HEDGES *Clare Leighton*
LETTERS FROM SKOKHOLM *R.M. Lockley*
THE UNOFFICIAL COUNTRYSIDE *Richard Mabey*
RING OF BRIGHT WATER *Gavin Maxwell*
THE SOUTH COUNTRY *Edward Thomas*
SALAR THE SALMON *Henry Williamson*

Also Available

THE LOCAL *Edward Ardizzone & Maurice Gorham*
A long out-of-print celebration of London's pubs
by one of Britain's most-loved illustrators.

LITTLE TOLLER BOOKS
Stanbridge Wimborne Minster Dorset BH21 4JD
Telephone: 01258 840549
ltb@dovecotepress.com
www.dovecotepress.com